Principal Component Analysis Handbook

Principal Component Analysis Handbook

Edited by **Rebecca Cross**

LANRYE
INTERNATIONAL

New Jersey

Published by Clanrye International,
55 Van Reypen Street,
Jersey City, NJ 07306, USA
www.clanryeinternational.com

Principal Component Analysis Handbook
Edited by Rebecca Cross

International Standard Book Number: 978-1-63240-416-9 (Hardback)

Printed in the United States of America.

Contents

Preface

This book on Principal component analysis (PCA) is a significant contribution to the field of data analysis. PCA involves a statistical procedure which orthogonally transforms a set of possibly correlated observations into set of values of linearly uncorrelated variables called principal components. The aim of this book is to enhance knowledge of scientists, engineers and researchers regarding the advantages of this technique in data analysis and includes information on the uses of PCA in distinct fields like multi-sensor data fusion, ecology, energy, agriculture, climate, image and video processing, gas chromatographic examination, color coating, materials science and automatic target identification.

All of the data presented henceforth, was collaborated in the wake of recent advancements in the field. The aim of this book is to present the diversified developments from across the globe in a comprehensible manner. The opinions expressed in each chapter belong solely to the contributing authors. Their interpretations of the topics are the integral part of this book, which I have carefully compiled for a better understanding of the readers.

At the end, I would like to thank all those who dedicated their time and efforts for the successful completion of this book. I also wish to convey my gratitude towards my friends and family who supported me at every step.

Editor

Principal Component Analysis –
A Realization of Classification Success
in Multi Sensor Data Fusion

Maz Jamilah Masnan, Ammar Zakaria, Ali Yeon Md. Shakaff,
Nor Idayu Mahat, Hashibah Hamid, Norazian Subari
and Junita Mohamad Saleh
Universiti Malaysia Perlis, Universiti Utara Malaysia & Universiti Sains Malaysia
Malaysia

1. Introduction

The field of measurement technology in the sensors domain is rapidly changing due to the availability of statistical tools to handle many variables simultaneously. The phenomenon has led to a change in the approach of generating dataset from sensors. Nowadays, multiple sensors, or more specifically multi sensor data fusion (MSDF) are more favourable than a single sensor due to significant advantages over single source data and has better presentation of real cases. MSDF is an evolving technique related to the problem for combining data systematically from one or multiple (and possibly diverse) sensors in order to make inferences about a physical event, activity or situation. Mitchell (2007) defined MSDF as the theory, techniques, and tools which are used for combining sensor data, or data derived from sensory data into a common representational format. The definition also includes multiple measurements produced at different time instants by a single sensor as described by (Smith & Erickson, 1991).

Although the concept of MSDF was first introduced in the 1960s and implemented in the 1970s in the robotic and defense application, today, the application of MSDF has proliferated into various nonmilitary applications. However the method is still disparate, where it is impossible to create a one-fits-all data fusion framework. The applications of MSDF are now multidisciplinary in nature. Some specific applications of MSDF include multimodal biometric systems using face and palm-print (Raghavendra et al., 2011); renewable energy system (Li et al., 2010); color texture analysis (Wu et al., 2007); face and voice outdoor multi-biometric system (Vajaria et al., 2007); medical decision making (Harper, 2005); image recognition (Sun et al., 2005), road traffic accidents (Sohn et al., 2003); and personal authentication (Duc et al., 1997; Kumar et al., 2006).

MSDF technique has become as a prominent tool in food quality assessment. Quality assessment in food processing industries aims to guarantee the standard and safety control of food products. Traditional approach of exploiting trained human panels to evaluate quality parameters can be replaced by artificial sensors. An example of artificial sensor receiving great interest from researcher in these industries is the electronic nose (i.e. e-nose) sensor that mimics the function of human smell. In the context of MSDF, usually e-nose is applied with another sensor called electronic tongue (i.e. e-tongue) which imitates the human taste function. Several applications of e-nose and e-tongue in food research include flavor sensing system (Cole et al., 2011); honey classification (Zakaria et al., 2011); classification of *orthosiphon stamineus* (Zakaria et al., 2010); detection of polluted food (Maciejak et al., 2003); discrimination of standard fruit solutions (Boilot et al., 2003); quality control of yoghurt fermentation (Cimander et al., 2002); and discrimination of several types of fruit juices (Winquist et al., 1999).

It is believed that the application of MSDF such as the fusion of e-nose and e-tongue, may overcome some drawbacks of using trained human panels especially for on-line food production. The use of artificial sensors is capable of overcoming human exhaustion and stress, minimize between-panels variability, and obviously human panels are not suitable for online measurements. Thus, this chapter focuses on the application of Principal Component Analysis (PCA) and Linear Discriminant Analysis (LDA) in MSDF. Two models of MSDF proposed by Hall (1992) namely low level data fusion and intermediate level data fusion are proposed in order to identify and classify different types of pure honey, beet sugar, cane sugar and adulterated samples (i.e. mixtures of pure honey with cane sugar and beet sugar). This chapter also aim to provide a concept to the constructive and lists some advantageousness of PCA in the application of MSDF especially in the analysis of multivariate data.

1.1 The fusion of artificial sensors

The appreciation of food is basically based on the combination of many human senses including sight, touch, sound, taste and smell. However, due to the expensive cost of having panels of trained expert to evaluate food quality parameters, a more rapid technique for objective measurement of food products in a consistent and cost-effective manner is highly needed in the food industry (Winquist et al., 2003). Two human senses that are believed to be closely correlated in the perception of flavour are the sense of smell and taste. The e-nose and e-tongue have been defined as the artificial sensing systems capable of producing a digital fingerprint of a given chemical ambient (D'Amico, 2000). Both devices consist of chemical sensor arrays coupled with an appropriate pattern recognition system capable of extracting information from complex signals (Buratti et al., 2004).

Basically, an e-nose is formed by having an array of gas sensors with different selectivity, a signal collecting unit and suitable pattern recognition software, all controlled and executed by a computer. The principle of e-tongue is similar to that of the e-nose, except for the array of sensors designed for liquids (Cosio et al., 2007). The ultimate task of these sensors is to collect the digital fingerprint or signals that would be further interpreted using multivariate statistical tools before the objective of the fusion approach is attained. One of the most popular exploratory data analyses in chemical sensors is PCA (Di Natale et al., 2006). PCA is a procedure that permits to extract useful information from the data, to explore the data structure, the relationship between the objects and features, and the global correlation of the features. Further details of PCA are described in Section 2. The selected principal components based on certain criteria will be used as an input for classification procedure using linear discriminant analaysis (LDA). Further descriptions of this technique are illustrated in section 3 of this chapter.

The selected architecture of MSDF in this research focuses on the approach of identity fusion. Identity fusion is a fusion of parametric data to determine the identity of an observed object. Our interest is to convert multiple sensor observations of a target attributes (such as e-nose and e-tongue responses) to a joint declaration of target identity. One of the key issues in developing an MSDF system is to determine the stage or phase in the data flow to combine or fuse the data (Hall & Llinas, 1997). In an identity fusion, Hall (1992) suggested three frameworks to be applied; (i) low level data fusion (or data level fusion); (ii) intermediate level data fusion (or feature level fusion); and (iii) high level data fusion (or decision level fusion). However, for the purpose of this discussion only data level and feature level fusion are discussed.

1.1.1 Low level data fusion

In low level data fusion, the e-nose and e-tongue sensors observe the target objects independently, and later the raw sensor data (i.e. original data collected from each sensor) are combined. In order to fuse raw sensor data, the original sensor data must be commensurate i.e. must be observations of similar physical quantities (Hall et al., 1997). Sometimes, the number of features recorded by the e-nose and e-tongue are different, but the raw sensor data can still be fused if both datasets are of the same sample size (equal n). It is important to ensure the new dataset is formed from the original non-normalized data. A framework of low level data fusion is illustrated in Fig. 1.

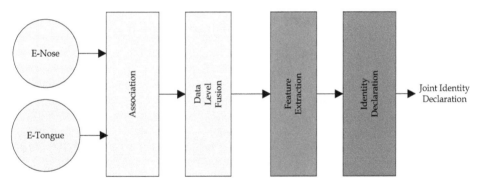

Fig. 1. Framework of low level data fusion by Hall (1992)

It is believed that the low level data fusion in identity fusion provides the most accurate result (Hall et al., 1997). This may be due to the fact that the originality information from each sensor is maintained and used in further processes. Thus, low level data fusion is potentially more accurate than the other two fusion methods. However, the difficulties in the application of low level data fusion method are due to the noise that frequently occurs in the sensor data and redundant data, which have an adverse effect on the classification results.

1.1.2 Intermediate level data fusion

This approach consists of extracting features from the signals of each sensor to yield feature vectors. Then, the feature vectors are fused and identity declaration is made based on the joint feature vectors. The identity declaration process includes techniques such as knowledge-based approach that includes expert system and fuzzy logic, or training-based approach like discriminant analysis, neural network, Bayesian technique, k-nearest neighbors and centre mobile algorithms. Fig. 2 illustrates the framework of the intermediate level data fusion.

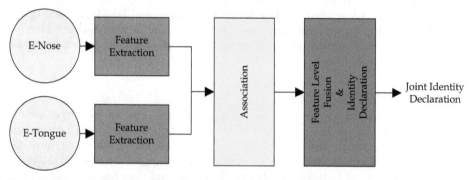

Fig. 2. Framework of intermediate level data fusion by Hall (1992)

It is important to note that both low and intermediate level data fusion apply feature extraction in transforming the raw signals provided by the sensor into a reduced vector of features describing parsimoniously the original information. Then, in the identity declaration, a quality class is assigned to the signals based on the feature extraction result.

2. Principal component analysis

Principal component analysis (PCA) was first described by Karl Pearson in 1901. A description of practical computing methods came much later from Harold Hotelling in 1933 (Manly, 2004). The idea of PCA is to keep the variation of the number of p original features into a fewer number of k unobservable variables ($k \leq p$), which is termed as principal components, as maximum as possible. Let Table 1 below describes the original data of a sensor data set with n objects each was observed with p features.

Case	X_1	X_2	$\cdots$	X_p
1	X_{11}	X_{12}	.	X_{1p}
2	X_{21}	X_{22}	.	X_{2p}
.	.	.	.	.
.	.	.	.	.
.	.	.	.	.
n	X_{n1}	X_{n2}	$\cdots$	X_{np}

Table 1. The form of data for a principal component analysis with p features on n cases

The aim of PCA is to find a new set of variables, say $Z_1, Z_2, ..., Z_i$ in a form of a linear combination of X's which is $Z = \alpha^T X$ .Here, $Z = (Z_1, Z_2, .., Z_p)$ is a vector of principal components and α^T is a matrix of coefficients α_{ij} for $i, j = 1, 2, .., p$.

The first principal component (Z_1) is the linear combination of the original features which mathematically written as $Z_1 = \alpha_{11}X_1 + \alpha_{12}X_2 + ... + \alpha_{1p}X_p$, assemble as the largest as possible of variance of p features subject to the condition that $\alpha_{11}^2 + \alpha_{12}^2 + ... + \alpha_{1p}^2 = 1$. Then, the second principal component (Z_2) is chosen to have the property of having the second largest possible variance of $X_1, X_2, ..., X_p$ while being uncorrelated with the first component (Z_1). The remaining principal components are defined similarly, with the jth principal component having the largest possible variance given that it is uncorrelated with the ith principal component for $i < j$. Let λ_i be the variance (eigenvalues) of Z_i, and α_{ij} be the eigenvectors of Z_i where $i, j = 1, 2, \cdots, p$, then these conditions hold for the eigenvalues and eigenvectors:

$$\lambda_1 \geq \lambda_2 \geq ... \geq \lambda_i \geq 0 \tag{1}$$

$$\alpha_i^T \alpha_i = 1 \tag{2}$$

$$\alpha_i^T \alpha_h = 0 \quad \text{where } i \neq j \tag{3}$$

Before we proceed to discuss on the issue of reducing the dimension intended for further analysis, it is a need to understand which matrix of information should be used, either a correlation matrix or a covariance matrix to allow for a computation of principal components. One should clearly understand when to use either one of the input matrix as often the results of these two are different. The next sections 2.1 and 2.2 briefly discuss the guidelines.

2.1 Information matrix for principal component analysis

2.1.1 Principal component using covariance matrix

An implicit assumption when using covariance matrix as an input is that the features should not have grossly different variances. Such differences in variance might arise because of different scales of measurements, different magnitude of measurements, or some combination of the two factors (Krzanowski, 2000). If they do, then the first few principal components will be pulled toward those features with the larger variances (Dillon & Goldstein, 1984).

In such cases, the data should be standardized and it means the correlation matrix is used in the PCA. As a general guideline, it would seem sensible to standardize first whenever the measured features show differences in variances, or whenever the user is concerned with very different measured entities or units (Krzanowski, 2000). However, transformation on the original data would result PC scores of a different meaning (Martinez & A.R. Martinez, 2001). Obviously, the big drawback of PCA based on covariance matrix is the sensitivity of the PCs to the units of measurement used for each element of X (Jolliffe, 2002).

2.1.2 Principal component analysis using correlation matrix

PCA aims to create linear combination of new variables that are uncorrelated to each other, thus, if the correlation matrix portrays nearly small correlation, then there is probably not much point in carrying PCA (Chatfield & Collins, 1980). PCA calculation based on correlation matrix is suitable for features with unequal scales of measure. One way to trace unequal scales is through wide differing variances among the features. In computing a correlation coefficient between two features, differences due to the mean and the dispersion of the features are removed (Dillon & Goldstein, 1984). This is recommended as the original features are all standardized to unit variance (Borgognone et al., 2001).

Therefore, data that is used to calculate PCA for correlation input does not need any transformation as it is applied automatically in the correlation computation. However, a disadvantage in using correlation matrix to calculate the principal components are that they give coefficients for standardized variables and are therefore less easy to interpret directly. Thus, to interpret the principal components in terms of the original variables, each coefficient must be divided by the standard deviation of the corresponding variables (Jolliffe, 2002).

2.2 Deciding the number of components to retain

Mathematically, the choice of values for coefficients α is subjected to the restrictions given in equations (2) and (3). Thus, the obtained principal components are in decreasing order of variance, $\text{var}(Z_1) \geq \text{var}(Z_2) \geq ... \geq \text{var}(Z_p) = \lambda_1 \geq \lambda_2 \geq ... \geq \lambda_p$. In practice, only the first k numbers of principal components account for most of the variability of the original data, thus keeping all the p principal components sound impractical. This mean, only the first k principal components will be used in further analysis while the p-k principal components will be ignored. However, there is no universally accepted method to do so because the decision is largely judgemental and a matter of taste (Dillon & Goldstein, 1984). A number of procedures to determine k have been suggested. Among the most common procedures are as follows.

2.2.1 Average eigenvalue

The most common criterion to determine the number of informative principal components in PCA is the Guttman-Kaiser criterion (Jackson, 1993). Principal components associated with eigenvalues (λ) derived from a covariance matrix which are larger in magnitude than the average of the eigenvalues, are retained. In the case of eigenvalues derived from a correlation matrix, the average is 1.0 for the variables to retain. Therefore, any principal component associated with an eigenvalue whose magnitude is greater than or equal to 1.0 is choosen for further analysis. However, Rencher (1998) warned that this method works well in practice but when it identifies wrongly, it is likely to retain too many components. It is well known as simple and the most suitable criterion to be applied especially when confronted with numerous variables.

2.2.2 Proportion of total variance explained

In a PCA model, each eigenvalue represents the level of variation of the original features explained by the associated principal components. Another popular decision criterion is based on the proportion of the total variance explained by the principal components retained in the model. If k-components are retained, then we may represent the cumulative variance explained by the first k principal components by,

$$t_k = \frac{\sum_{i=1}^{k} \lambda_i}{\sum_{i=1}^{p} \lambda_i} \tag{4}$$

Often, the researcher decides on a satisfactory value for t_k and then determines k accordingly. The obvious problem with the technique is to decide on an appropriate t_k. In practice, it is common to select from 70% to 90% (Jolliffe, 2002). Because of such obviously arbitrary, this approach has sometimes been criticized for its subjectivity (Kim & Mueller, 1978). While Jackson (1993) strongly argues against the use of this method except possibly for exploratory purposes when little are known about the population of the data.

2.2.3 Scree plot

Perhaps much easier decision on k can be made based on graphical approaches as suggested by Cattell (1966) called the scree plot. A scree plot is a plot of the eigenvalues versus the index of the eigenvalue. With this approach, the eigenvalues of each component are plotted in successive order of their extraction, and then an elbow in the curve is identified by applying a straightedge to the bottom portion of the eigenvalues to see where they form an approximate straight line (Dillon & Goldstein, 1984).

The value of k is given by the point at which the components curve above the straight line formed by the smaller eigenvalues. Fig. 3 shows a case in which k is equal to three and the straight line (shallow) begins at the forth until the last component. As we can observe from Fig. 3, the third component is marked exactly at eigenvalue is equal to 1. Dillon and Goldstein (1984) argue that this method is inconclusive when there is no obvious break or there may be several breaks. And it become more troublesome when two breaks occur among the first half of the eigenvalues, since it will be difficult to decide which of the breaks reflect the correct number of components.

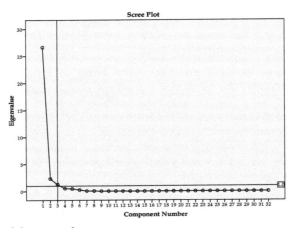

Fig. 3. Illustration of the scree plot.

3. Linear discriminant analysis

Linear discriminant analysis or discriminant function analysis or in short discriminant analysis is a supervised technique for classifying objects into two or more groups, given the measurements for these objects is from several features (i.e. sensor responses). It involves deriving linear combinations of the independent features that will discriminate between the a *priori* defined groups in such a way that the misclassification error are minimized (Dillon & Goldstein, 1984). The discrimination can be accomplished by maximizing the between group variance relative to the within-group variance. The basic discriminant analysis is the one that involves only two-group problem which was first suggested by R. A. Fisher (1936). In the two-group problem, the aim is to find a single linear composite of the predictor features that could discriminate between the two groups. The linear composite then acts as a new axis along which the groups were maximally separated.

In reality, we may encounter discrimination problems of more than two groups which require an extension of the basic discriminant analysis called the multiple discriminant analysis. The goal in multiple discriminant analysis is much similar with discriminant analysis for two groups. Dillon and Goldstein (1984) describe in general, with k groups and p predictor features, there are in total, min(p, k-1) possible discriminant functions (i.e. linear composites). In most applications, since the number of features (p) is exceeding the number of groups (k), at most k-1 discriminant functions will be considered. However, not all of these functions show statistically significant variation among the groups, and fewer than k-1 discriminant functions is actually needed. Likewise in forming principal components in PCA, discriminant functions are generated so that the scores of each new discriminant function are uncorrelated with the scores of previously obtained discriminant function. Thus, each linear composite is the new single function that maximizes the ratio of the between-groups to within-groups variability, accordingly. Besides, the discriminant functions are extracted in a decreasing order of accounted variation.

There are assumptions that need to be considered by researchers for obtaining optimal procedure in the sense of producing smallest misclassification error rate. According to Dillon and Goldstein (1984), for optimality, we assume (i) multivariate normality of the p predictor features, and (ii) equal variance-covariance matrices in each of the k groups. They added that the objectives of multiple discriminant analysis are for the most part is the generalizations of those of the two-group problem. Among others it includes:

i.   To find the linear composites with as large as possible between-groups variability subject to each uncovered linear composites being uncorrelated with previously extracted composites. The accounted variations for all linear composites are in decreasing order.
ii.   To determine whether the group centroids are statistically different.
iii. To determine the number of discriminant functions that is statistically significant.
iv.   To successfully assign new signal or observation to one of the several groups.
v.   To determine the predictor features that contributes most for discrimination among groups.

The goal in constructing classification rules is to minimize the mistakes in assigning new signals to its groups. Less mistakes means less error for the classification rules to correctly allocate the signals. In real problem, often one has a set of data to be discriminated accordingly to g groups. However, using the same data for constructing a rule and evaluating a rule is biased. As the matter of fact, it does not mimic the real use of discrimination rule to classify a future object where the rule is constructed based on the existing data. There are some techniques that can be considered in an attempt to avoid such bias. Some of the techniques are re-substitution method, cross validation method which is also known as sample-splitting method and leave-one-out method. Lachenbruch and Mickey (1968) in (Krzanowski, 2000) proposed the leave-one-out method that was believed to be able to overcome most problems inherent in the previous two methods. The technique consists of determining the allocation rule using the sample data minus one observation and then using the subsequent rule to classify the omitted observation. Repeating this procedure by omitting each of the individuals in the two training set in turn yields, an estimate of the error rates, the proportions of misclassified signals in the two training sets.

4. Materials and methods

The experiment was implemented in the Sensor Laboratory, Centre of Excellence for Advanced Sensor Technology, University Malaysia Perlis. The aim is to identify and classify different types of pure honey, beet sugar, cane sugar and adulterated samples (i.e. mixtures of pure honey with cane sugar and beet sugar) by applying the low level data fusion and intermediate level data fusion. PCA was employed to reduce the data dimension and further classification was fulfilled by LDA.

4.1 Sample selection and preparation

In this experiment, 10 different brands of *Tualang* honey were purchased from the local market with three different batches of each particular honey. While for the adulteration purposes, two types of sugar solution namely beet sugar and cane sugar were imported from Germany and United Kingdom respectively. Display of pure honey and sugar are illustrated in Fig. 4 and all honey and sugar samples are summarized in Table 2.

Item	Descriptions	Group
AG	Agromas	1
AS	As-Syifa	1
ST	Syair Timur	1
T3	Tualang 3	1
TB	Tayyibah	1
TK	Tualang King	1
TLH	Tualang TLH	1
TN	Tualang Napis	1
WT	Wild Tualang	1
YB	Yubalam Bahtera	1
BS	Beet Sugar	2
CS	Cane Sugar	3
XXBS	Pure Honey + BS	4 (20%), 5 (40%)
XXCS	Pure Honey + CS	6 (20%), 7 (40%)

Table 2. Description and abbreviation of honey samples, sugar and adulterated samples used in the experiments

Based on the three different batches of each pure honey, three samples of 5ml was prepared for further measurement. For adulteration samples, each pure honey was mixed with sugar of different concentration (i.e. 20% and 40%) as shown in Table 3. Each pure sugar was also measured. Each sampling of pure honey, sugar and adulterated were repeated ten times. In total there were about 172 samples of pure honey, pure sugar and adulterated mixtures.

Percentage of Pure Honey	Descriptions
20% pure honey	1:4 (ratio of pure honey / sugar solution)
40% pure honey	2:3 (ratio of pure honey / sugar solution)

Table 3. Description of mixture for different samples of honey and sugar

Fig. 4. Display of different samples of honey and sugar

4.2 Electronic nose setup and measurement

The e-nose used was Cyranose320 from Smith Detection™, consists of 32 non-selective sensors of different types of polymer matrix, blended with carbon black composite, configured as an array. It can be trained to analyze both simple and complex vapor mixtures with equal ease. When the sensors are exposed to vapors or aromatic volatile compounds they swell, changing the conductivity of the carbon pathways and causing an increase in the resistance value that is monitored as the sensor signal. The resistance changes across the array are captured as a digital pattern i.e. representative of the test smell (Dutta et al., 2006).

The e-nose setup for this experiment is illustrated in Fig. 5 and the setting of the sniffing cycle is also indicated in Table 4. Each sample was drawn from the bottle using 10ml syringe and kept in a 13 x 100 mm test tube and seal with a silicone stopper. Each sample was replicated ten times. Before measurement, each sample was placed in a heater block and heat up for 10 minutes to generate sufficient headspace volatiles. The temperature of sample was controlled at $50 \pm$ °C during the headspace collection.

Preliminary experiments were performed to determine the optimal experimental setup for the purging, baseline purge and sample draw durations. Ten seconds baseline purge with 40 seconds sample draw produced an optimal result (result is not shown). Baseline purge was set longer to ensure residual gases were properly removed since all the samples are in a liquid form and contains moisture. The pump setting was set to medium speed during sample draw. The filter used is made up of activated carbon granules and has large surface area which is effective to remove a wide range of volatile organic compounds and moisture in the ambient air. The experiment was carried out using e-nose for a variety of honey samples followed by sugar and adulterated samples.

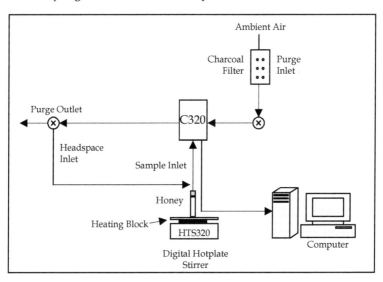

Fig. 5. E-nose setup for headspace evaluation of honey, sugar concentration and adulteration sample

	Cycle	Time (s)	Pump Speed
Sampling	Baseline Purge	10	120 mL/min
Setting	Sample Draw	40	120 mL/min
	Idle Time	3	-
	Air Intake Purge	40	120mL/min

Table 4. E-nose parameter setting for honey, sugar and adulterated samples assessment

4.3 Electronic tongue setup and measurement

The chalcogenide-based potentiometric e-tongue was made up of eleven distinct ion-selective sensors from Sensor Systems (St. Petersburg, Russia). The e-tongue system shown in Figure 6 was implemented by arranging an array of potentiometric sensors around the reference probe. Table 5 describes the potentiometric sensors used in this experiment. Each sensor output was connected to the analogue input of a data acquisition board (NI USB-6008) from National Instruments (Austin TX, USA).

A 10% (w/v) solution of honey in distilled water was prepared and stirred for 3 minutes at 1000rpm before making any measurements. Each sample was replicated ten times. For each measurement, the e-tongue was steeped simultaneously and left for two minutes, and the potential readings were recorded for the whole duration. After each sampling, the e-tongue was rinsed twice using distilled water (stirred at 400rpm for two minutes) to remove any sticky residues from previous sample sticking on the sensor surface to avoid contaminating of the next sample.

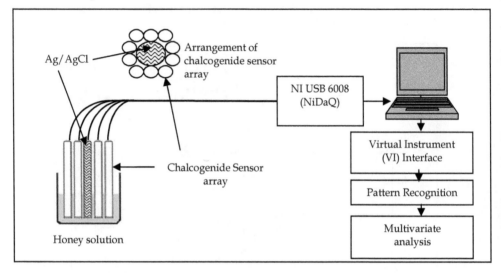

Fig. 6. E-tongue setup for headspace evaluation of honey, sugar concentration and adulterated sample

Sensor Label	Description
Fe^{3+}	Ion-selective sensor for Iron ions
Cd^{2+}	Ion-selective sensor for Cadmium ions
Cu^{2+}	Ion-selective sensor for Copper ions
Hg^{2+}	Ion-selective sensor for Mercury ions
Ti^+	Ion-selective sensor for Titanium ions
S^{2-}	Ion-selective sensor for Sulfur ions
$Cr(VI)$	Ion-selective sensor for Chromium ions
Ag^+	Ion-selective sensor for Argentum ions
Pb^{2+}	Ion-selective sensor for Plumbum ions
HI 5311	pH sensor
HI 2111	Reference probe using Ag/AgCl electrode

Table 5. Chalcogenide-based potentiometric electrodes used in the e-tongue.

4.4 Data preprocessing

The fractional measurement method is essential when using a multi-modalities sensor fusion. This technique is often known as baseline manipulation and was applied to preprocess the data of both modalities (Gardner & Bartlett, 1999). The maximum sensor response, S_t is subtracted from the baseline, S_0 and then divided again by the S_0. The formula for this dimensionless and normalized S_{frac} is determined as follows:

$$S_{frac} = [S_t - S_0]/S_0 \qquad (5)$$

This gives a unit response for each sensor array output with respect to the baseline, which compensates for sensors that have intrinsically large varying response levels. It can also further minimize the effect of temperature, humidity and temporal drifts (Gardner & Bartlett, 1999).

The data from different modalities were processed separately and all sensors were used in this analysis. In the case of the e-nose, S_0 is the minimum value taken during the baseline purge with ambient air and S_t was measured during the sample draw. Each sampling cycle was repeated three times and the average was obtained for each of ten replicated samples. For the e-tongue measurements, S_0 (baseline reading) is the average reading of distilled water, while S_t is the sensor reading when steeped in the solution. The steeping cycle was repeated three times for each sample and the average was obtained for each ten of the replicated samples. Each S_{frac} data point from each e-nose and e-tongue sensor formed the S_{frac} matrix for further analyses.

4.4.1 Low level data fusion

For the purpose of low level data fusion, measurements recorded from both sensors were fused during the data level. For the e-nose data, there were 720 observations with 32 features from 16 different honey, sugar and adulterated samples. Likewise for the e-tongue data, 720 observations with 11 features from 16 different honey, sugar and adulterated samples were recorded. As a result, a new dimension for the fused data was represented by 720 observations with 43 features. At this stage, the original data from both measurements is formed in a data matrix, and is described in Fig. 7 as follows. No transformation is being applied at this stage.

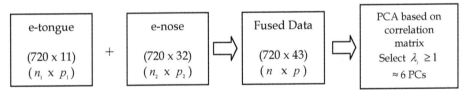

Fig. 7. Illustration of fusing data in low level data fusion

The correlation input matrix from the fused data was proceeded for the PCA calculation. For the purpose of classification in LDA, the reduced number of principal components was selected based on magnitude eigenvalues greater or equal to 1 ($\lambda_i \geq 1$). The result from the scree plot is also applied for comparison and confirmation purposes.

4.4.2 Intermediate level data fusion

In this framework, fusion was applied after feature extraction process. For that purpose, PCA was calculated based on the correlation matrix from both datasets. The number of principal components to retain is decided based on the associated eigenvalues with magnitude greater than or equal to 1.0 ($\lambda_i \geq 1$). The results were double checked using the scree plot of each dataset. Fig. 8 illustrates the related processes. The resulting principal components from each sensor which is three principal components were then combined before the classification using LDA is performed.

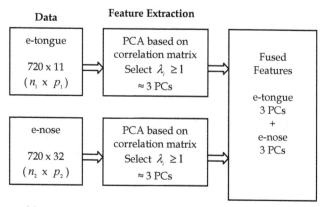

Fig. 8. Illustration of fusing extracted features in intermediate level data fusion

5. Results and discussion

Before the analyses of PCA was continued, a thorough study on each and every selected principal components (i.e. at low level data fusion) considered for classification using LDA was performed and the resulting classification error rate for each case are highlighted in Fig. 9. Comparisons and evaluations of classification error rate were performed differently based on correlation or covariance input matrix, procedure to evaluate performance of leave-one-out approach and the elimination of the least important of principal components (i.e. elimination begin with principal components of the smallest eigenvalue). Table 6 shows the total of variance explained using the correlation and covariance matrix input for the low level data fusion.

Number of Retained Principal Component	Correlation Matrix		Covariance Matrix	
	Total Eigenvalue (%)	Error Rate (Leave-one-out)	Total Eigenvalue (%)	Error Rate (Leave-one-out)
2	72.070	0.546	77.277	0.546
3	77.836	0.500	86.640	0.500
⋮	⋮	⋮	⋮	⋮
20	99.581	0.151	99.999	0.153
21	99.649	0.151	100.000	0.150
⋮	⋮	⋮	⋮	⋮
42	99.998	0.142	100.000	0.142
43	100.000	0.144	100.000	0.144

Table 6. Total variance explained for low level data fusion

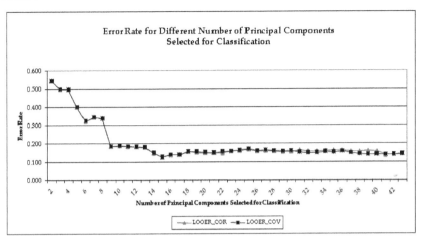

Fig. 9. Different classification performance for correlation and covariance input matrix with leave-one-out approach.

Fig. 9 clearly reveals similar classification performance of correlation and covariance input matrix with a leave-one-out approach for the low level data fusion. It should be highlighted here that the performance of classification for the correlation and covariance input is not much differ because the standard deviations for each features in the fused dataset is slightly small.

In reality, good classification performance is not determined by the greater number of features included in data. What we need is features with the most discriminative effect which often measured by the error rate. In the case of low level data fusion, the PCA based on the correlation matrix of fused data was used to extract the most important features in a linear combination form. Table 7 displays the total of variance explained for the principal components of low level data fusion. Six principal components with eigenvalues greater than or equal to 1.0 were retained to be the input for classification using LDA. It can be seen that with only six linear combinations of the original features out of 43-principal-component, we only loose about 9.3% of information to proceed with classification task. The scree plot in Fig. 10 also shows that six principal components should be retained.

Component	Extraction Sums of Squared Loadings		
	Total	% of Variance	Cumulative %
1	27.167	63.179	63.179
2	3.823	8.891	72.070
3	2.480	5.767	77.836
4	2.223	5.169	83.005
5	1.966	4.572	87.577
6	1.356	3.153	90.730
⋮	⋮	⋮	⋮

Table 7. Total variance explained for low level data fusion

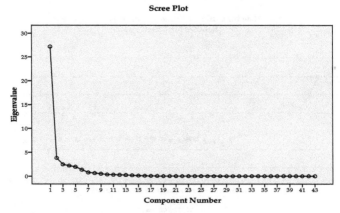

Fig. 10. Scree plot for the low level data fusion

Component	Extraction Sums of Squared Loadings		
	Total	% of Variance	Cumulative %
1	4.023	36.573	36.573
2	2.232	20.289	56.862
3	1.930	17.549	74.411
⋮	⋮	⋮	⋮

Table 8. Total variance explained of e-tongue data for intermediate level data fusion

Component	Extraction Sums of Squared Loadings		
	Total	% of Variance	Cumulative %
1	26.652	83.287	83.287
2	2.336	7.300	90.587
3	1.287	4.023	94.610
⋮	⋮	⋮	⋮

Table 9. Total variance explained of e-nose data for intermediate level data fusion

Table 8 and 9 display the total of variance explained for the principal components of intermediate level data fusion. Based on the eigenvalues greater than or equal to 1.0 from both e-tongue and e-nose data, three principal components each were retained to be the input for classification using LDA. With the three principal components selected from e-tongue and e-nose data, we loose about 31% of information which is quite high compared to the low level data fusion. The scree plot in Fig. 11 seems agrees that three principal components are adequate to represent the original features.

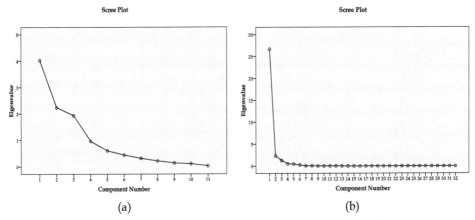

Fig. 11. Scree plot for (a) e-tongue data and (b) e-nose data low level data fusion

The selected principal components for low and intermediate level data fusion are further analyzed. The classification and prediction of the class of different types of pure honey, sugar, and adulterated samples were carried out using LDA with leave-one-out procedure. Table 10 indicates the significant differences in means of the predictors (i.e. the selected principal components) between the seven groups for both fused models. The results indirectly show the importance of the principal component to the discrimination function. Based on the Wilk's Lambda, principal component with smaller value means it is an important predictor. The most important principal components to the least important were arranged according to the italic number. Note in contrast, the bigger the Wilk's Lambda, the smaller the F values. Besides knowing the important predictors for the discrimination function, it is worth to investigate whether the assumption of homogeneity of covariance matrices is met. Table 11 displays the Box's M test for both data fusion models. The significant values of both data fusion models indicate that the covariance matrices are not similar for the seven groups.

Tests of Equality of Group Means						
Low Level Data Fusion			Intermediate Level Data Fusion			
	Wilks' Lambda	F	Sig.		Wilks' Lambda	F	Sig.
PC1	.777⁵	34.109	.000	PC1_EN	.794⁵	30.742	.000
PC2	.686²	54.404	.000	PC2_EN	.612²	75.467	.000
PC3	.741⁴	41.578	.000	PC3_EN	.928⁶	9.206	.000
PC4	.739³	42.005	.000	PC1_ET	.718⁴	46.707	.000
PC5	.399¹	178.960	.000	PC2_ET	.676³	56.940	.000
PC6	.921⁶	10.183	.000	PC3_ET	.423¹	162.029	.000

Table 10. Test of equality of group means to identify the important variable to the discrimination function

Test Results					
Low Level			Intermediate Level		
Box's M		3194.447	Box's M		3450.654
F	Approx.	22.505	F	Approx.	24.310
	df1	126		df1	126
	df2	6677.884		df2	6677.884
	Sig.	.000		Sig.	.000

Table 11. Test null hypothesis of equal population covariance matrices.

Based on Table 12 and Table 13, all the first five discriminant functions for low and intermediate level data fusion are able to explain 100% of the total variance. However, the canonical correlation values greater than 0.5 reveal that only the first two discriminant functions from both fusion model describe strong relationship.

Eigenvalues				
Function	Eigenvalue	% of Variance	Cumulative %	Canonical Correlation
1	7.151	75.1	75.1	.937
2	2.177	22.9	98.0	.828
3	.106	1.1	99.1	.309
4	.076	.8	99.9	.267
5	.008	.1	100.0	.090
6	.000	.0	100.0	.000

Table 12. Percentage of variance explained for each discrimination function for low level data fusion.

Eigenvalues				
Function	Eigenvalue	% of Variance	Cumulative %	Canonical Correlation
1	6.365	74.3	74.3	.930
2	2.015	23.5	97.8	.818
3	.105	1.2	99.1	.309
4	.074	.9	99.9	.263
5	.006	.1	100.0	.077
6	.000	.0	100.0	.002

Table 13. Percentage of variance explained for each discrimination function for intermediate level data fusion.

The best predictors in predicting the types of honey, sugar, and adulterated samples from the respective discrimination functions of each data fusion model are marked italic in Table 14. The highest value in each function (column) marks as the best predictor. For example, the best predictor for the first discriminant function of the low level data fusion is the third principal components (PC3).

Standardized Canonical Discriminant Function Coefficients						
Function (Low Level Data Fusion)						
	1	2	3	4	5	6
PC1	-.420	.758	.277	.583	.404	-.165
PC2	-.342	.902	.353	-.500	-.018	.265
PC3	1.299	-.128	.139	.275	.057	.661
PC4	1.097	-.452	.352	-.312	.497	-.281
PC5	-1.236	-.580	.171	-.008	.179	.218
PC6	.115	-.317	.716	.193	-.591	-.207
Standardized Canonical Discriminant Function Coefficients (cont'd)						
Function (Intermediate Level Data Fusion)						
	1	2	3	4	5	6
PC1_T	1.238	-.117	.161	.340	-.089	.628
PC2_T	.968	-.363	.384	-.144	.621	-.166
PC3_T	-1.263	-.512	.146	.074	.177	.258
PC1_N	-.084	.615	-.153	.562	.636	-.027
PC2_N	-.020	.923	.434	-.320	.119	.275
PC3_N	.005	.032	.741	.551	-.287	-.297

Table 14. Indication of relative importance of the independent variables in predicting the groups for both data fusion models.

Graphical representations of the classification for low level data fusion and intermediate level data fusion are as of Fig. 12 and Fig. 13 respectively. Table 15 and 16 describes in detail the classification results for each fusion model. It seems that the classification of several types of pure honey (group 1), beet sugar (group 2) and cane sugar (group 3) were very good. Confusions occur a lot for adulterated samples of group 4, 5, 6 and 7. As we can see the classification performance of the intermediate level data fusion based on the leave-one-out approach is slightly better than the classification performance of the same approach of low level data fusion with 73.5% and 71.5% correct classification respectively.

	Cross-validated Classification Results of Leave-One-Out Procedure							
Group	Predicted Group Membership							Total
	1	2	3	4	5	6	7	
Count 1	300	0	0	0	0	0	0	300
2	0	10	0	0	0	0	0	10
3	0	0	10	0	0	0	0	10
4	6	10	4	49	0	31	0	100
5	0	0	8	0	41	4	47	100
6	1	6	0	35	0	58	0	100
7	0	0	0	0	53	0	47	100

Table 15. Classification performance for low level data fusion

	Cross-validated Classification Results of Leave-One-Out Procedure							
Group	Predicted Group Membership							
	1	2	3	4	5	6	7	Total
Count 1	300	0	0	0	0	0	0	300
2	0	10	0	0	0	0	0	10
3	0	0	10	0	0	0	0	10
4	7	9	4	46	0	34	0	100
5	0	0	10	1	45	5	39	100
6	2	6	0	23	4	65	0	100
7	0	0	0	0	47	0	53	100

Table 16. Classification performance for intermediate level data fusion

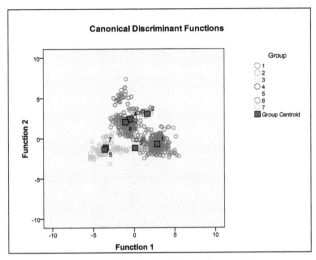

Fig. 12. Seven groups discriminating plot for low level data fusion

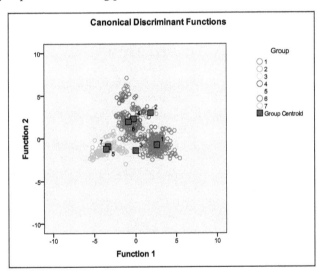

Fig. 13. Seven groups discriminating plot for intermediate level data fusion

6. Conclusions

This study focuses on the application of PCA in reducing the dimension of fused data from e-tongue and e-nose at low level and intermediate level data fusion. Previous studies on PCA have proven that this method is strongly advisable to be applied before performing any classification. In this study, we have shown the ability of PCA to create new variables in the form of principal components of the original features. Even though with some loss of information, special characteristics preserved in the selected principal components have made the new variables as reliable predictors in the discrimination and classification process. In order to improve the classification performance of the multi sensor data fusion models in this study, there are two special attentions that should be given. Firstly, to fulfil the discriminant analysis assumption on the homogeneity of covariance for each group, and secondly to study and overcome the violation effect to discriminant analysis method caused by the existence of outliers. In future, we will attempt to solve these problems.

7. Acknowledgement

The equipment used in this project was provided by the Universiti Malaysia Perlis (UniMAP). This project is also funded by the Fundamental Research Grants Scheme (9003-00250), Ministry of Higher Education Malaysia (MOHE) and Short Term Grant (2011), Universiti Sains Malaysia (USM). The authors take this opportunity to express their sincere gratitude to Prof. Mohd Noor Ahmad (UniMAP), and Assoc. Prof. Abdul Hamid Adom (UniMAP) for their support. The authors acknowledge the financial sponsorship provided by UniMAP and MOHE, under the Academic Staff Training Scheme.

8. References

Afifi, A., A. Clark, V., & May, S. (2004). *Computer-Aided Multivariate Analysis*. Chapman & Hall, ISBN 1-58488-308-1, Boca Raton, Florida

Berrueta, L.A. & Alonso-Slaces, R.M., Heberger, K. (2007). Supervised pattern recognition in food analysis. *J. Chrom: A*, Vol. 1158, pp. 196-214

Boilot, P.; Hines, E. L.; Gongora, M.A. & Folland, R. S. (2003). Electronic Noses Inter-Comparison, Data Fusion and Sensor Selection in Discrimination of Standard Fruit Solutions. *Sensors and Actuators*, Vol. B 88, pp. 80-88

Borgognone, M. G.; Bussi, J. & Hough, G. (2001). Principal Component Analysis: Covariance or Correlation Matrix. *Food Quality and Preference*, Vol. 12, pp. 323-326

Buratti, S.; Benedetti, S.; Scampicchio, M. & Pangerod, E. C., (2004). Characterization and Classification of Italian Barbera Wines by Using an Electronic Nose and an Amperometric Electronic Tongue. *Analytica Chimica Acta*, Vol. 525, September 2004, pp. 133-139

Cattell, R. B. (1966). The scree test for the number of factors. *Multiv.Behav. Res.*, Vol. 1, pp. 245-276.

Chatfield, C. & Collins, A. J. (1980). *Introduction to multivariate analysis*; Chapman and Hall, ISBN 0-412-16030-7, Great Britain

Cimander, C.; Carlsson, M. & Mandenius, C. (2002). Sensor Fusion for On-Line Monitoring of Yoghurt Fermentation. *Journal of Biotechnology*, Vol. 99, pp. 237-248.

Cole, M.; Covington, J. A. & Gardner, J. W. (2011). Combined Electronic Nose and Electronic Tongue for a Flavor Sensing System. *Sensors and Actuators B: Chemical*, Vol. 156, Issue 2, pp. 832-839

Cosio, M. S.; Ballbio, D.; Benedetti, S. & Gigliotti, C. (2007). Evaluation of Different Conditions of Extra Virgin Olive Oils with an Innovative Recognition Tool Built by Means of Electronic Nose and Electronic Tongue. *Food Chemistry*, Vol. 101, February 2006, pp. 485-491

D'Amico, A.; Di Natale, C. & Paolesse, R. (2000). Portraits of Gasses and Liquids by Arrays of Nonspecific Chemical Sensors: Trends and Perspectives. *Sensors and Actuators B*, Vol. 68, 2000, pp. 324-330

Di Natale, C.; Martinelli, E.; Pennazza, G.; Orsini, A. & Santonico, M. (2006). Data Analysis for Chemical Sensor Array. *Advances in Sensing with Security Applications*, pp.147-169.

Dillon, W. R. & Goldstein, M. (1984). *Multivariate analysis, methods and applications*. John Wiley & Sons, Inc., ISBN 0-471-08317- 8, New York, USA

Duc, B.; Bigun, E. S.; Bigun, J.; Maitre, G. & Fischer, S. (1997). Fusion of Audio and Video Information for Multi Modal Person Authentication. *Pattern Recognition letters*, Vol. 18, pp. 835-843

Dutta, R.; Das, A.; Stocks, N.A.; Morgan, D. (2006). Stochastic Resonance-based Electronic Nose: A Novel Way to Classify Bacteria. *Sensors and Actuators B*, Vol. 115, pp. 17-27

Gardner, J.W. & Bartlett, P.N. (1999). Electronic Noses: Principals and Applications. Oxford University Press: Oxford, 0-19-855955-0, UK

Gnanadesikan, R. (1997). *Methods for statistical data analysis of multivariate observations*. John Wiley and Sons, Inc., ISBN 0-471-16119-5, New Jersey, USA

Hall, D. L. (1992). *Mathematical techniques in Multisensor Data Fusion*. Artec House Inc., ISBN 0-89006-558-6, Boston, London

Hall., D. L. & Llinas, J. (1997). An Introduction to Multisensor Data Fusion. *Proceedings of the IEEE*, *58*, 6-22

Hardle, W. & Simar, L. (2007). *Applied multivariate statistical analysis;* Springer, ISBN 3-540-72243-4, Berlin

Harper, P. R. (2005). A Review and Comparison of Classification Algorithms for Medical Decision Making. *Health Policy*, Vol. 71, pp. 315-331

Jackson, D.A. (1993). Stopping Rules in Components Analysis: A Comparison of Heuristical and Statistical Approaches. *Ecology*, Vol. 74, pp. 2204–2214

Jolliffe, I. T. (2002). *Principal Component Analysis*. 2nd. Ed. Springer, ISBN 0-387-95442-2, New York, USA

Kim, J. O. & Mueller, C. W. (1978). *Factor Analysis: Statistical Methods and Practical Issues*. Sage, ISBN 9780803911666, Beverly Hill, CA

Krzanowski, W. J. (2000). *Principal of Multivariate Analysis, A User's Perspective*. Oxford, ISBN 0-19-850708-9, New York, USA

Kumar, A.; Wong, D. C. M.; Shen, H. C. & Jain, A. K. (2006). Personal Authentication using Hand Images. *Pattern Recognition Letters*, Vol. 27, pp. 1478-1486

Li, J.; Luo, S. & Jin, J. S. (2010). Sensor Data Fusion for Acurate Cloud Presence Prediction using Dempster-Shafer Evidence Theory. *Sensors*, Vol. 10, pp. 9384-9396.

Maciejak, T. R.; Kukawska-Tarnawska, B.; Tyszkiewicz, J. & Tyszkiewicz, S. (2003). Multi-Sensor Odour Detection and Measurement of Polluted Food. *Polish Journal of Food and nutrition Sciences*, Vol. 12/53, pp. 45-48

Manly, B. F. J. (2004). *Multivariate Statistical Methods: a Primer*. Chapman & Hall, ISBN 1-58488-414-2, Boca Raton, Florida

Martinez, W. L. & Martinez, A. R. (2001). *Computational Statistics Handbook with Matlab*. Chapman & Hall/CRC, ISBN 1-58488-229-8, London, UK

Mitchell, H.B. (2007). *Multi-Sensor Data Fusion, an Introduction*. Springer, ISBN 978-3-540-71463-7, Heidelberg, Berlin

Persaud, K.; Dodd, G. (1982). Analysis of discrimination mechanisms in the mammalian olfactory system using a model nose. *Nature*, Vol. 299, pp. 352-355

Raghavendra, R.; Dorizzi, B.; Rao, A., & Kumar, G. H. (2011). Designing Efficient Fusion Schemes for Multimodal Biometric Systems using Face and Palmprint. *Pattern Recognition*, Vol. 44, pp. 1076-1088

Rencher, A. C. (1998). *Multivariate Statistical Inference and Applications*. Wiley, ISBN 0-471-57151-2, New York

Smith, C. R. & Erickson, G. J. (1991). Multisensor Data Fusion: Concepts and Principals. *IEEE Pacific Rim Conference on Communications, Computers and Signal Processing*, pp. 235-237

Sohn, S. Y. & Lee, S. H. (2003). Data Fusion, Ensemble and Clustering to Improve the Classification Accuracy for the Severity of Road Traffic Accidents in Korea. *Safety Science*, Vol. 41, pp. 1-14

Sun, Q.; Zeng, S.; Liu, Y.; Heng, P. & Xia, D. (2005). A New Method of Feature Fusion and its Application in Image Recognition. *Pattern Recognition*, Vol. 38, pp. 2437-2448

Vajaria, H.; Islam, T.; Mohanty, P.; Sarkar, S.; Sarkar, R. & Kasturi, R. (2007). Evaluation and Analysis of a Face and Voice Outdoor Multi-Biometric System. *Pattern Recognition Letters*, Vol. 28, pp. 1572-1580

Winquist, F.; Krantz-Rülcker, C. & Lundström, I., (2003). Electronic Tongues and Combinations of Artificial Senses, In: *Sensors Update*, Vol. II, Baltes, H.; Fedder, G. K. & Korvink, J. G., pp. 279-306, Wiley-VCH, ISBN 3-527-30601-3, Germany

Winquist, F.; Lundström, I. & Wide, P. (1999). The Combination of an Electronic Tongue and Electronic Nose. *Sensors and Actuators B*, Vol. 58, pp. 512-517

Wu, Y.; Li, M. & Liao, G. (2007). Multiple Features Data Fusion Method in Color Texture Analysis. *Applied Mathematics and Computation*, Vol. 185, pp. 784-797

Zakaria, A.; Shakaff, A. Y. M.; Adom, A. H.; Ahmad, M. N.; Masnan, M. J.; Aziz, A. H. A.; Fikri, N. A.; Abdullah, A. H. & Kamarudin, L. M. (2010). Improved Classification of *Orthosiphon stamineus* by Data Fusion of Electronic Nose and Tongue Sensors, *Sensors*, Vol. 10, pp. 8782-8796, ISSN 1424-8220

Zakaria, A.; Shakaff, A. Y. M.; Masnan, M.J.; Ahmad, M. N.; Adom, A. H.; Jaafar, M. N.; A. Ghani, S., Abdullah, A. H.; Aziz, A. H. A.; Kamarudin, L. M.; Subari, N. & Fikri, N. A. (2011). A Biomimetic Sensor for the Classification of Honeys of Different Floral Origin and the Detection of Adulteration. *Sensors*, Vol. 11, pp. 7799-7822, ISSN 1424-8220

Methodology for Optimization of Polymer Blends Composition

Alessandra Martins Coelho[1], Vania Vieira Estrela[2],
Joaquim Teixeira de Assis[3] and Gil de Carvalho[3]
[1]Instituto Federal de Educacao, Ciencia e Tecnologia do Sudeste de Minas Gerais
(IF SUDESTE MG), Rio Pomba, MG,
[2]Departamento de Telecomunicacoes, Universidade Federal Fluminense (UFF), Niterói, RJ,
[3]Instituto Politécnico (IPRJ), Universidade Estadual do Rio de Janeiro (UERJ),
Nova Friburgo, RJ,
Brazil

1. Introduction

The research of polymer blends, or alloys, has experienced enormous growth in size and sophistication in terms of its scientific base, technology and commercial development (Paul & Bucknall, 2000). As a consequence two very important issues arise: the increased availability of new materials and the need for materials with better performance.

Polymer blends are polymer systems originated from the physical mixture of two or more polymers and/or copolymers, without a high degree of chemical reactions between them. To be considered a blend, the compounds should have a concentration above 2% in mass of the second component (Hage & Pessan, 2001; Ihm & White, 1996). However, the commercial viability of new polymers has begun to become increasingly difficult, due to several factors.

The advantages of polymer blends lie in the ability to combine existing polymers into new compositions obtaining in this way, materials with specific properties. This strategy allows for savings in research and development of new materials with equivalent properties, as well as versatility, simplicity, relatively low cost (Koning et al., 1998) and faster development time of new materials (Silva, 2011).

Rossini (2005) mentions that economically and environmentally, a very viable alternative is to replace the recycling of pure polymers by mixtures of discarded materials. Mechanical recycling causes the breakdown of polymer chains, which impairs the properties of polymers. This degradation is directly proportional to the number of cycles of recycling. Therefore, the blend of two or more discarded polymers can be a realistic alternative, since it can result in materials with very interesting properties, at a low cost. Besides its inexpensiveness, this choice is also a smart solution to the reutilization of garbage. Post-consumption package disposal always occurs in a disorderly manner and without regard for the environment. The recycling process becomes increasingly more important and necessary to remediate environmental impact.

According Pang et al. (2000) apud Marconcini & Ruvolo Filho (2006) polyolefins such as high density polyethylene (HDPE), low density polyethylene (LDPE) and polypropylene (PP) and polyesters such as poly (ethylene terephthalate) (PET) are classes of thermoplastics that have been widely used in packaging and constitute a large part of post-consumer waste. The recycling of these materials and their mechanical characterization anticipating the possibility of a new cycle of life in the form of new products is challenging, although technologically and environmentally correct (Marconcini & Ruvolo Filho, 2006).

The polymer blends can be obtained basically in two ways (Rossini, 2005):

- By dissolving the polymers in a good solvent, common to them, and subsequently letting the solvent evaporate; and
- In a mixer where the working temperature is high enough to melt or mollify the polymeric components, without causing degradation of the same.

According to Wessler (2007), the polymer blends may be miscible or immiscible. The miscibility is the most important property to be analyzed in a blend, given that all other system properties depend on the number of phases, their morphology and adhesion between them. The miscibility term is directly related to the solubility, i.e., a blend is miscible when the polymers dissolve in each other mutually (Silva, 2011). The immiscible between the various engineering polymers is a limiting factor for its production. Thus, it is necessary to use compatibilization agents for their production.

Computational modeling has become increasingly popular. The main objective of models is to assist process optimization with minimal investment of time and resources for experimental work. Most techniques are classified into two main groups: physical models and statistical models as shown by Malinov & Sha (2003).

Statistical methods are chosen according to research objectives. There are several multivariate analysis methods for purposes quite different from each other. The desired value and quality of one or more product characteristics can be obtained via experiment analysis and DOE. These methods help determining optimal settings and controllable factors of a process such as: temperature, pressure, amount of reagents, operating time, etc.. When compared to the method of trial and error, DOE also allows a reduction of the number of required tests, and savings in time, labor and money.

An important application of DOE is the optimization of experimental formulations as, for example, the composition of mixtures. The formulation development is a fundamental part of the food industry, chemicals, plastics, rubber, paints, medicines, and the like.

In materials science, it is important to understand the correlation between material processing, microstructure and properties that enable the optimization of process parameters and compositions of materials to achieve the desired combination of properties, according Malinov & Sha (2003).

The problem presented here is to determine the fraction of each polymer blend component, and to determine the agent or, in some cases, an agents system, when it is necessary to use more than one compatibilizing agent. Thus, this text studies the effect of factors, for example, amount of polypropylene, additive type, and amount of additive in the composition of polymer blends, i.e., the optimal polymer blends formulation using factorial design.

Pawlak *et al.* (2002) pointed out that the elongation at break and impact strength of recycled HDPE/PET blends has increased with the addition of EGMA or maleic anhydride grafted styrene-ethylene butylene-styrene (SEBS-*g*-MA). The best results were obtained for PET/HDPE/EGMA at 75%/25%/4 pph and PET/HDPE/SEBS-*g*-MA at 75%/25%/10 pph. The mechanical properties of the blends were related to the phase dispersion. The increase in the viscosities of the compatibilized blends was observed due to the reaction during blending. Carvalho et al. (2003) considered blend composition complexity as a function of the ideal percentage of each one of their components in their computer study for optimization of polymeric blends. With the objective of analyzing the mechanical behavior of the blend in relation to PET and to PP, the same speed test was adopted for the three tested materials. The results are presented in Table 1.

	Modulus of Elasticity [MPa]	Tensile Strength at Break [MPa]	Elongation at Rupture [%]
PET	2230	50.2	3.2
PET/PP 75/25	1740	31.3	17
PP	1130	26.9	615

Table 1. Results of the traction for PET, PP and the blend PET/PP.

2. Design of Experiments (DOE)

2.1 Introduction to design of experiments

One of the most common and challenging problems in experiments concerns the determination of the influence that one or more variables has on the variable of interest. Designed experiments address these problems and also have extensive application in the development of new processes and design of new products. Some of its applications are

- Characterization of a process (experiment screening): It aims to determine which factors affect the response;
- Optimization of an experiment: It aims to determine the important factors in the region leading to an optimal response; and
- Product planning: It tries to determine the factors that influence the most the verification effort.

A DOE is the pre-requisite for a successful experimental study (Tang et al., 2010). Assuming that the goal of experimentation is to find a function, or at least a satisfactory approximation of it, which acts on k factors producing observed responses (as outlined in Figure 1), the system acts like an initially unknown transfer (or modifying) function, which operates on the factors, producing as output, the observed responses. Thus, a better understanding of the nature of the reaction under study in order to choose the best system operating conditions (Silva, 2011).

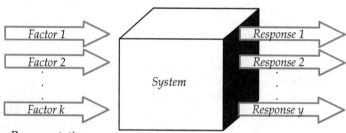

Fig. 1. System Representation

2.2 Factorials design

In a designed experiment, the data-producing process is actively manipulated to improve the data quality and to eliminate redundancy. A common goal of all experimental designs is to collect data as parsimoniously as possible while providing sufficient information to accurately estimate model parameters. By factorial experiment we mean that in each replication of the experiment, all possible combinations of levels are investigated.

Multilevel designs is used to systematically vary experimental factors and then assign each factor a discrete set of *levels*. Full factorial designs (FD) measure response variables using every treatment (combination of the factor levels).

Plackett-Burman designs are used when only main effects are considered significant. They require a number of experimental runs that are a multiple of 4 rather than a power of 2.

Binary factor levels are indicated by ±1. The design is for eight runs manipulating seven two-level factors. The number of runs is a fraction $8/2^7 = 0.0625$ of the runs required by a full factorial design. Economy is achieved at the expense of confounding main effects with any two-way interactions.

2.2.1 Two-level designs

Two-Level designs are often used in experiments involving several factors, in which is necessary to study the combined effect of factors on a response. However, several special cases of general factorial design are important because they are widely used in research and form the basis for other designs of considerable practical value. The most important of these special cases is of k factors, where each one has only two levels.

When planning an experiment, one should first determine the factors and the answers adequate to the system under study. The factors, that is, the variables controlled by the experimenter, can be both quantitative (such as values of temperature, pressure or time) and qualitative (such as two machines, two operators, levels "up" and "down" of a factor, or perhaps the presence or absence of a factor). Depending on the problem, there may be more than a response of interest and, eventually, these responses can also be qualitative.

After determination of the factors to be observed, it is necessary to implement the factorial design, i.e., the values of the factors that will be used in the experiment. All possible combinations of factors are investigated. Among the many advantages of factorial design, the following (Button, 2005) can be named:

a.   The number of trials can be reduced without jeopardizing the quality of information;
b.   It permits simultaneous study of several variables while separating its effects;
c.   It assesses the reliability of results;
d.   It allows stepwise research realization which in general adds new tests an iterative; and
e.   It selects the variables that influence a process with a minimum number of tests;

In factorial design, the factors and levels are pre-determined by setting and they correspond to a fixed effects model. This type of planning is normally used in the early stages of research. Since there are only two levels for each factor analysis, its assumed that the response variable presents a linear behavior between these levels (Button, 2005). Effects are defined as "the change in response down level (-) for the up level (+)" and they can be classified in two categories: main effect (effect on the level change of a single factor) and interaction effect (effect on the change in level between two or more factors at the same time).

2.2.2 2^2 factorial design

Geometrically, the design 2^2 can be represented by a square where each vertex corresponds to an experiment.

Figure 2 shows, geometrically, the 2^2 factorial design and its planning matrix. The letters A and B represent the factors. The levels are represented by - and +, which correspond to low and high levels of factors. The combination of experiments, with both factors at low level is represented by the number 1. The effects of interest in the 2^2 factorial design are the main effects A (represented by number 2) and B (represented by number 3). The interaction factor AB, also called contrast (represented by the number 4) is generated from the product of the signs of the columns of the main effects A and B.

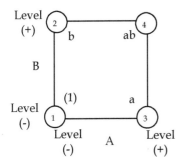

Treatment Combination	Effects		Responses
	A	B	
(1)	-	-	y_1
2	+	-	y_2
3	-	+	y_3
4	+	+	y_4

Fig. 2. Geometrical Notation and Planning Matrix for 2^2 Factorial Design.

The main effect of A is by definition the average of the effects of A in two levels of B. The same happens with the main effect B, as seen in (1) and (2).

$$A = y_+ - y_- = (\frac{y_2 + y_4}{2}) - (\frac{y_1 + y_3}{2}) \tag{1}$$

$$B = y_+ - y_- = (\frac{y_4 + y_3}{2}) - (\frac{y_2 + y_1}{2}) \qquad (2)$$

The interaction effect AB is given by:

$$AB = (\frac{y_1 + y_4}{2}) - (\frac{y_2 + y_3}{2}) \qquad (3)$$

2.2.3 2^3 factorial design

2^3 factorial designs have three factors at two different levels, which request the performance of eight experimental trials (each of these experiments in which the system is subjected to a defined set of levels). Based on factors that you want to study and their levels, it is possible to build a planning matrix as shown in Table 1. The first column of the effects (A factor) is filled alternating one by one the levels of factors (- + - + ...), column 2 (B factor) is filled alternating two by two the levels of factors (- - + + ...) and, finally, the third column (C factor) the first four experiments are filled with the lowest level and last four with the higher level (- - - - + + + +). The combination of experiments with both factors at low level (-) is also represented by the number 1.

Based on the planning matrix (Table 2) it is possible to generate the table of contrast coefficients. This matrix is composed of three main effects (A, B and C) and four interaction effects (AB, AC, BC and ABC). Table 3 shows the signs of effects for the 2^3 factorial design.

Treatment	Effects		
Combination	A	B	C
(1)	-	-	-
2	+	-	-
3	-	+	-
4	+	+	-
5	-	-	+
6	+	-	+
7	-	+	+
8	+	+	+

Table 2. Planning Matrix 2^3 Factorial Design

In conformity to Neto et al. (2003), the effects on the 2^3 factorial design can also be interpreted as contrasts geometric, whose representation is a cube, in which the eight trials of the planning matrix corresponding to its vertices. The main effects and interactions of two factors are contrasts between two planes, which can be identified by examining the coefficients of contrast. In general, one main effect on the planning 2^3 is a contrast between the opposite sides and perpendicular to the axis of the corresponding variable. The interactions between two factors, in turn, are contrasts between two diagonal planes. These planes are perpendicular to a third plane, defined by the axes of the two variables involved in the interaction.

Treatment Combination	Effects							
	I	A	B	C	AB	AC	BC	ABC
(1)	+	-	-	-	+	+	+	-
2	+	+	-	-	-	-	+	+
3	+	-	+	-	-	+	-	+
4	+	+	+	-	+	-	-	-
5	+	-	-	+	+	-	-	+
6	+	+	-	+	-	+	-	-
7	+	-	+	+	-	-	+	-
8	+	+	+	+	+	+	+	+

Table 3. Signs of Effects for the 2^3 Factorial Design.

If K is the number factors, then a general form for the effects can be given by:

$$ef = \frac{1}{2^{k-1}} X^T y \text{ , and}$$ (4)

$$M_{ef} = \frac{1}{2^k} X^T y \text{ .}$$ (5)

2.2.4 Fractional designs

For experiments with many factors, two-level full FD can lead to large amounts of data. For example, a two-level full factorial design with 11 factors requires $2^{11} = 2048$ runs. Often, however, individual factors or their interactions have no distinguishable effects on a response. This is especially true of higher order interactions. As a result, a well-designed experiment can use fewer runs for estimating model parameters.

Fractional FD use a fraction of the runs required by full FD. A subset of experimental treatments is selected based on an evaluation (or assumption) of which factors and interactions have the most significant effects. Once this selection is made, the experimental design must separate these effects. In particular, significant effects should not be confounded, that is, the measurement of one should not depend on the measurement of another. The challenge is to choose basic factors and generators so that the design achieves a specified resolution in a specified number of runs. The confounding pattern shows that main effects are effectively separated by the design, but two-way interactions are confounded with various other two-way interactions.

2.3 Response Surface Methodology (RSM)

RSM is defined how a collection of mathematical and statistical techniques useful for the modeling and analysis of problems in which a response of interest is influenced by several process variables (termed factors) whose objective is to optimize this response (Montgomery, 2005; Box & Draper 1987; Myers & Montgomery, 1995 apud Tang et al., 2010).

Box & Draper (1987) define RSM how a collection of statistical techniques useful in researches, with the purpose to determine the best conditions and give greater insight into the nature of certain phenomena. It comprises the following three main components (Tang et al., 2010):

a. Experimental design to determine the process factors values based on which the experiments are conducted and data are collected;
b. Empirical modeling to approximate the relationship (i.e. the response surface) between responses and factors; and
c. Optimization to find the best response value based on the empirical model.

It can be assumed that the system under study is governed by a function which is described by the experimental variables. Normally this function can be approximated by a polynomial, which provides a good description of the factors and response. The order of the polynomial is limited by the type of planning used. Two-level FD, fractional or complete, can only estimate main effects and interactions. Factorial design with three levels (central point) can estimate, moreover, degree of curvature in the response. In general, the relationship is:

$$y = f(x_1, x_2, \cdots, x_k) + \varepsilon, \tag{6}$$

where the true response f is unknown and sometimes very complicated; ε represents disturbances in f, such as, measurement error on the response, background noise, the effect of other variables, and so on. In any planned experiment, there is a strong relationship between the analysis of a designed experiment and a regression analysis that can be used for predictions of an experiment 2^k.

Because f is unknown, we must approximate it. In fact, successful use of RSM is critically dependent upon the experimenter's ability to develop a suitable approximation for f. Usually, a low-order polynomial is sought after.

The first-order model is likely to be appropriate when the experimenter is interested in approximating the true response surface over a relatively small region of the independent variable space in a location where there is little curvature in f.

To describe these models in a screening study, are used simple polynomials, i.e., those containing only linear terms. A simple model of a response y in an experiment with two controlled factors x_1 and x_2, two polynomials is:

$$y = \beta_0 + \beta_1 x_1 + \beta_2 x_2 + \varepsilon \tag{7}$$

$$y = \beta_0 + \beta_1 x_1 + \beta_2 x_2 + \beta_{12} x_1 x_2 + \varepsilon, \tag{8}$$

where x_1 and x_2 are main effects; $x_1 x_2$ is a two-way interaction effect; β_0 is the average value of all responses; ε includes both experimental error and the effects of any uncontrolled factors in the experiment; and β_1, β_2 and β_2, are, respectively, the coefficients related to the main variables x_1 and x_2, and the coefficient for the interaction between x_1 and x_2. So, x_1 and x_2 should be manipulated while measuring y, with the objective of accurately estimating $\beta 0$, $\beta 1$ and $\beta 2$. Equations (7) and (8) can be combined and the resulting model is given by:

$$\hat{y} = X\beta , \tag{9}$$

where $\hat{y}$ is the vector of responses estimated by model; X is the coefficient contrast matrix; and β is the coefficient of the model or regression vector. In RSM design, there should be at least three levels for each factor. In this way, the factor values that are not actually tested using fewer experimental combinations and the combinations themselves can be estimated (Neseli et al., 2011). The effect of a factor is defined as the variation in the response produced by the change in the factor level.

3. Development and discussion

Factorial DOE has been used to measure the influence of the following input variables: amount of polypropylene, additive type and amount of additive on the values of response variables. Relevant mechanical properties for polymeric blends PET/PP are ME, elongation at rupture and TS at rupture. The following experiments were accomplished by a 2^3 factorial design. Their specifications are presented in the Table 4.

	PET	PP
Manufacturing	Fairway	Polibrasil
Type	201050 NT	TM 6100
Apparent density [g/m3] ASTM-D 1505	0.88	0.5
Index of fluidity [g/10 min] ASTM-D 1238	(*)	16
Intrinsic viscosity [dl/g]	0.82	(*)
Melting [ºC] ASTM-D 3418	> 240	160 - 175

(*) = not available

Table 4. Specification supplied by the manufacturers of PET and PP (Carvalho et al., 2003).

The factors will be analyzed on two levels (top and bottom) according to data presented in Table 5.

Main Effects	Factors	Level (-)	Level (+)
A	Amount of polypropylene	5%	25%
B	Additive type	C2 (acrylic acid)	C1 (maleic anhydride)
C	Amount of additive	1%	5%

Table 5. Planning Matrix

The preparation of test specimens and tests were performed according to the Standard Test Method for Tensile Properties of Plastics - ASTM D-638 (2010). The mechanical properties of ME, elongation at rupture and TS were evaluated in ten executions for each test.

Tables of contrast coefficients for ME (Table 6), contrast coefficients for study of Strain at Break (Table 7) and contrast coefficients for TS (Table 8) were obtained from the Table 3 and Table 4. All tables were composed by three main effects: A (amount of polypropylene), B (additive type), C (amount of additive), and the four interaction effects AB, AC, BC and ABC. The last column of each table contains the values of Y_n (n = 1, 2 and 3, respectively,

ME, elongation at rupture and TS at rupture), corresponds to the average of the experimental results found for each test, in 10 executions.

Treatment Combination	Effects								Y_1 (MPa)
	I	A	B	C	AB	AC	BC	ABC	
(1)	+	-	-	-	+	+	+	-	1605
2	+	+	-	-	-	-	+	+	1448
3	+	-	+	-	-	+	-	+	1445
4	+	+	+	-	+	-	-	-	1371
5	+	-	-	+	+	-	-	+	1562
6	+	+	-	+	-	+	-	-	1355
7	+	-	+	+	-	-	+	-	1550
8	+	+	+	+	+	+	+	+	1232

Table 6. Contrast Coefficients and average values by modulus of elasticity.

Treatment Combination	Effects								Y_2 (%)
	I	A	B	C	AB	AC	BC	ABC	
(1)	+	-	-	-	+	+	+	-	4.36
2	+	+	-	-	-	-	+	+	3.80
3	+	-	+	-	-	+	-	+	4.01
4	+	+	+	-	+	-	-	-	3.60
5	+	-	-	+	+	-	-	+	4.22
6	+	+	-	+	-	+	-	-	4.55
7	+	-	+	+	-	-	+	-	4.50
8	+	+	+	+	+	+	+	+	4.24

Table 7. Contrast Coefficients and average values by elongation at rupture

Treatment Combination	Effects								Y_3 (MPa)
	I	A	B	C	AB	AC	BC	ABC	
1	+	-	-	-	+	+	+	-	50
2	+	+	-	-	-	-	+	+	41
3	+	-	+	-	-	+	-	+	43
4	+	+	+	-	+	-	-	-	37
5	+	-	-	+	+	-	-	+	46
6	+	+	-	+	-	+	-	-	40
7	+	-	+	+	-	-	+	-	48
8	+	+	+	+	+	+	+	+	37

Table 8. Contrast Coefficients and average values by tensile strength at rupture

4. Calculation of effects and results interpretation

The 8 x 8 matrix factorial design is

$$X = \begin{bmatrix} +1 & -1 & -1 & -1 & +1 & +1 & +1 & -1 \\ +1 & +1 & -1 & -1 & -1 & -1 & +1 & -1 \\ +1 & -1 & +1 & -1 & -1 & +1 & -1 & +1 \\ +1 & +1 & +1 & -1 & +1 & -1 & -1 & -1 \\ +1 & -1 & -1 & +1 & +1 & -1 & -1 & +1 \\ +1 & +1 & -1 & +1 & -1 & +1 & -1 & -1 \\ +1 & -1 & +1 & +1 & -1 & -1 & +1 & -1 \\ +1 & +1 & +1 & +1 & +1 & +1 & +1 & +1 \end{bmatrix} \qquad (10)$$

Tables 6, 7 and 8 include all necessary values for calculating the effects on Modulus of Elasticity (ME), Strain at Break and TS. The column vectors Y_1, Y_2 and Y_3, with respective average values are shown in (11) and the product of X^T (1) by the respective vectors (11) appears in (12).

$$Y_1 = \begin{bmatrix} 1605 \\ 1448 \\ 1445 \\ 1371 \\ 1562 \\ 1355 \\ 1550 \\ 1232 \end{bmatrix} ; \quad Y_2 = \begin{bmatrix} 4.36 \\ 3.80 \\ 4.01 \\ 3.60 \\ 4.22 \\ 4.55 \\ 4.50 \\ 4.24 \end{bmatrix} ; \quad Y_3 = \begin{bmatrix} 50 \\ 41 \\ 43 \\ 37 \\ 46 \\ 40 \\ 48 \\ 37 \end{bmatrix} \qquad (11)$$

Returning to Tables 5, 6 and 7 can be seen that in all columns except the first, have four positive and four negative signs. To find the global average to fairly apportion the first element of each of the vectors $X^T.Y_1$, $X^T.Y_2$ e $X^T.Y_3$ by 8. The other elements of the vectors correspond to the effects and will be divided by 4, result in (13).

$$X^T Y_1 = \begin{bmatrix} 11568 \\ -756 \\ -372 \\ -170 \\ -28 \\ -294 \\ 102 \\ 194 \end{bmatrix} ; \quad X^T Y_2 = \begin{bmatrix} 33.28 \\ -0.9 \\ -0.58 \\ 1.74 \\ -0.44 \\ 1.04 \\ 0.52 \\ -0.74 \end{bmatrix} ; \quad X^T Y_3 = \begin{bmatrix} 342 \\ -32 \\ -12 \\ 0 \\ -2 \\ -2 \\ 10 \\ -8 \end{bmatrix} \qquad (12)$$

$$
\begin{bmatrix} \overline{Y_1} \\ A \\ B \\ C \\ AB \\ AC \\ BC \\ ABC \end{bmatrix} = \begin{bmatrix} 1446 \\ -189 \\ -93 \\ -42.5 \\ -7 \\ -73.5 \\ 25.5 \\ -48.5 \end{bmatrix} ; \quad \begin{bmatrix} \overline{Y_2} \\ A \\ B \\ C \\ AB \\ AC \\ BC \\ ABC \end{bmatrix} = \begin{bmatrix} 4.16 \\ -0.225 \\ -0.145 \\ 0.435 \\ -0.110 \\ 0.260 \\ 0.130 \\ -0.185 \end{bmatrix} ; \quad \begin{bmatrix} \overline{Y_3} \\ A \\ B \\ C \\ AB \\ AC \\ BC \\ ABC \end{bmatrix} = \begin{bmatrix} 42.75 \\ -8 \\ -3 \\ 0 \\ -0.5 \\ -0.5 \\ 2.5 \\ -2 \end{bmatrix} \quad (13)
$$

The Gauss method, which is a direct method for solving linear systems, can be used to solve the system found. In this case, the elements of columns of matrix X (10), that the corresponding effects were divided by 2, as shown in (14). The vectors y_1, y_2 and y_3 are the terms independent of the linear system. The results are the same as described in (13).

$$
X = \begin{bmatrix}
+1 & -\frac{1}{2} & -\frac{1}{2} & -\frac{1}{2} & \frac{1}{2} & \frac{1}{2} & \frac{1}{2} & -\frac{1}{2} \\
+1 & \frac{1}{2} & -\frac{1}{2} & -\frac{1}{2} & -\frac{1}{2} & -\frac{1}{2} & \frac{1}{2} & -\frac{1}{2} \\
+1 & -\frac{1}{2} & \frac{1}{2} & -\frac{1}{2} & -\frac{1}{2} & \frac{1}{2} & -\frac{1}{2} & \frac{1}{2} \\
+1 & \frac{1}{2} & \frac{1}{2} & -\frac{1}{2} & \frac{1}{2} & -\frac{1}{2} & -\frac{1}{2} & -\frac{1}{2} \\
+1 & -\frac{1}{2} & -\frac{1}{2} & \frac{1}{2} & \frac{1}{2} & -\frac{1}{2} & -\frac{1}{2} & \frac{1}{2} \\
+1 & \frac{1}{2} & -\frac{1}{2} & \frac{1}{2} & -\frac{1}{2} & \frac{1}{2} & -\frac{1}{2} & -\frac{1}{2} \\
+1 & -\frac{1}{2} & \frac{1}{2} & \frac{1}{2} & -\frac{1}{2} & -\frac{1}{2} & \frac{1}{2} & -\frac{1}{2} \\
+1 & \frac{1}{2} & \frac{1}{2} & \frac{1}{2} & \frac{1}{2} & \frac{1}{2} & \frac{1}{2} & \frac{1}{2}
\end{bmatrix} \quad (14)
$$

The three tables below show data contained in the vectors (13) in order to enable analysis of the influence of each factor individually and the interaction of these factors on the ME, strain at break and tensile strength (TS).

Table 9 shows that the three main effects, the factors of polypropylene amount, additive type and amount of additive reduce the ME. The amount of polypropylene is the major contributing factor to the reduction of elasticity. The model obtained for the ME is presented in (15).

Average:	1446
Main Effects:	
A (Amount of polypropylene)	-189
B (Additive type)	-93
C (Amount of Additive)	-42.5
Interaction between two factors:	
AB	-7
AC	-73.5
BC	25.5
Interaction between three factors:	
ABC	-48.5

Table 9. Effects calculated for the modulus elasticity.

$$modulus\ of\ elasticity = 1446 + 25.5 * B * C \qquad (15)$$

Figure 3 represents the RS for the ME as a function of B and C. The additive type and amount of additive increase the ME. Hence, the interaction between type and amount of additive can improve the interaction between molecules and compatibility of the mixture.

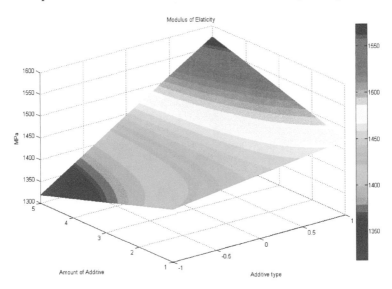

Fig. 3. Response surface for modulus of elasticity as a function of the factors B and C

With respect to the elongation at rupture, observed in Table 10, the main effect, amount of additive, increases the strain at rupture. The same happens with the interaction of two factors AC and BC. The obtained ME model appears in (16) and (17).

Average:	4.16
Main Effects:	
A (Amount of polypropylene)	-0.225
B (Additive type)	-0.145
C (Amount of Additive)	0.435
Interaction between two factors:	
AB	-0.110
AC	0.260
BC	0.130
Interaction between three factors:	
ABC	-0.185

Table 10. Effects calculated for elongation at rupture

$$elongation\ at\ rupture = 4.16 + 0.435 * C + 0.268 * A * C \qquad (16)$$

$$elongation\ at\ rupture = 4.16 + 0.435 * C + 0.13 * B * C \qquad (17)$$

Figure 4 represents the graphic of the response surface elongation at rupture as a function of the factors A and C. Note that the additive type and amount of additive increases the elongation at rupture, fact already observed in Table 10. Figure 4 show that this factor has a significant effect on elongation at rupture. It is evident in the Figures (4) and (5) that the amount of additive is more significant than the types of additive analyzed.

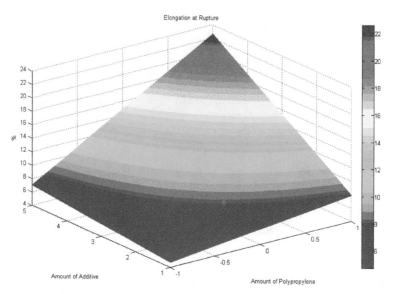

Fig. 4. Response surface for elongation at rupture as a function of the factors A and C

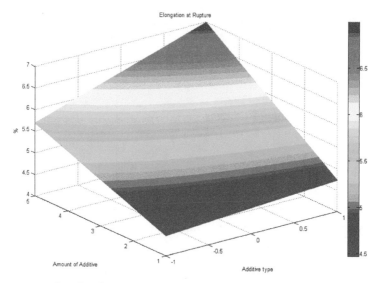

Fig. 5. Response surface for elongation at rupture as a function of the factors B and C

In Table 11, the main effect C has no significant value for TS, since the main effects A and B show a reduction. Interaction BC shows an increase in TS, while the interaction between the three factors (ABC) reduces TS. The model obtained for the modulus of TS is presented in (17).

Average:	42.75
Main Effects:	
A (Amount of polypropylene)	-8
B (Additive type)	-3
C (Amount of Additive)	0
Interaction between two factors:	
AB	-0.5
AC	-0.5
BC	2.5
Interaction between three factors:	
ABC	-2

Table 11. Effects calculated for tensile strength

Figure 6 represents the graphic of the response surface for TS as a function of the factors B and C. Note that the additive type and amount of additive increases the TS.

$$tensile\ strenght = 42.75 + 2.5 * B * C \qquad (18)$$

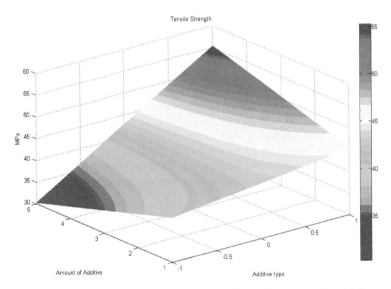

Fig. 6. Response surface for tensile strength as a function of the factors B and C

4.1 Geometrical interpretation of effects

The eight trials of each of the three planning matrices correspond to the vertices of the cube. The effects can be identified by examining the coefficients of contrast. Figure 5 reveals that the tests are all negative on one side of the cube, which is perpendicular to the axis of factor 1 (amount of polypropylene) and is located on the lower level of this factor. The other essays are on the opposite side, which corresponds to the upper level. The effect of factor 1 can be considered, therefore, as the contrast between these two faces of the cube. The effects, 2 and 3, also are contrasts between healthy opposite sides and perpendicular to the axis of the corresponding variable. The interaction between two factors, appear as contrasts between two diagonal planes. These planes are perpendicular to a third plane, defined by the axes of the two variables involved in the interaction.

Figure 7 presents the geometric interpretation of the effects. For instance, vertex 1 has the following coordinates: 5% polypropylene and 1% additive, which is acrylic acid.

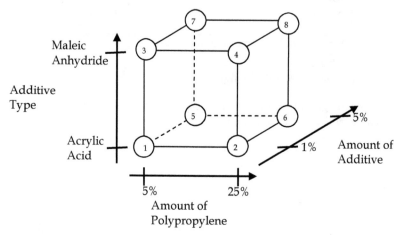

Fig. 7. Geometric interpretation of the effects

5. Model-based DOE (PCA-based DOE)

Nowadays, design, monitoring and optimization of applications by means of mathematical models are very advantageous in process control. Nevertheless, a trustworthy model that complies with operation constraints is as a rule difficult to develop not trivial. According to Asprey & Macchietto (2000), a wide-ranging modeling method comprises:

- An initial analysis and structure modeling of the system based on process knowledge;
- Designing optimal experiments according to the planned model;
- Perform experiments; and
- Using experimental information to estimate model parameters and accomplish model validation by probing available estimated parameters and existing data.

This chapter deals with experiments designed for a specific algebraic equations (AE) system called model-based DOE (MBDOE) while factorial analysis based on DOE uses empirical models. Numerical models are often nonlinear algebraic equations (NAE), dynamic algebraic equations (DAE), or partial differential equations (PDE). MBDOE is done before any real in order to describe structure selection, to model parameter estimation, and so forth. Pragmatically speaking, MBDOE sets up a DOE objective function.

From an algorithmic point of view, DOE has been combined with AE systems for a long time and applied to DAE systems by Zullo (1991) and Asprey & Macchietto (2002). Several optimal design criteria (ODC) have been suggested and considered by different case studies; Walter & Pronzato (1990) gave a detailed discussion of available ODC and their geometrical interpretations. Lately, Atkinson (2003) used DOE for non-constant measurement variance cases and Galvanin et al. (2007) extended the DOE territory to parallel experiment designs.

This work focuses on a DOE global methodology relying on PCA, so that a large system can separated into small pieces and a sequence of experiments can be designed to avoid numerical problems. Moreover, the problem can be transformed into familiar ODC under certain assumptions and a subset of model parameters can be chosen to boost estimation precision without changing the objective function form.

5.1 Parameter estimation

Parameter estimation can be generalized into the following optimization problem:

$$z = \min_{\theta} \sum_{i=1}^{n} \sum_{j=1}^{q} \left(y_{m,i,j} - f_j\left(t, x_i, \theta, u\right) \right)^2 , \tag{19}$$

subject to:

$$Hx = f_j\left(t, x_i, \theta, u\right)$$

$$u_{mim} \le u \le u_{max}$$

$$t_0 \le t \le t_f$$

$$x_{mim} \le x \le x_{max}$$

$$\theta_{mim} \le \theta \le \theta_{max}$$

where n is the number of experiments, q is the number of equations, respectively, y stands for measured variables and subscript m indicates a measurement. x is the state variables of the DAE system. For simplicity, the variables x are assumed to be measurable, thus $y=x.f$ represents the DAE equations and H is used to discriminate algebraic and dynamic equations (the corresponding rows for AEs are zero). θ stands for the model parameters and u has the controlled variables. Assume the control profile u is known over a predefined time interval $[t_0, t_f]$. In parameter estimation, the only unknown in integrating f is θ and normally the boundary of θ is defined according to the nature of the process to be modeled. The measurement noises is considered a multivariate normal distribution $(N(0, V_m))$, otherwise Eq. (19) needs to be rebuilt from MLE according to the specific noise distribution function. In most cases, normally distributed noise is a safe assumption. Eq. (19) is similar to the classical optimal control problem in which the objective function usually is $Z = \min_{u} x(t_f)$.

This dynamic system optimization problem can be solved by sequential and simultaneous methods.

In sequential approaches, only the unknown variables (e.g., θ for parameter estimation, u for optimal control) are discretized and manipulated directly by the non-linear programming (NLP) solver. After the unknown variables are updated, the DAE is integrated given the initial condition x_0 and integration interval $[t_0, t_f]$.

For simultaneous methods, the entries of x are discretized along t and approximated by polynomials between two neighboring discretization grids. Thus, the integration step is avoided and both state and unknown variables are changed by NLP directly with certain constraints. A review of these methods can be found in Espie & Macchietto (1989). After the NLP solver converges, the corresponding θ is our best estimate $(\hat{\theta})$ based on the measurements at hand. To evaluate the accuracy of the estimation, the posterior covariance matrix (parameter covariance matrix) is defined by:

$$V(\hat{\theta}, \varphi) = \left[\sum_{r=1}^{q} \sum_{s=1}^{q} v_{m,rs}^{-1} J_r J_s + V_0^{-1} \right]^{-1},$$ (20)

where φ is the design vector which typically contains the time, initial state condition, control variables, etc. $v_{m,rs}$ is the r-th term in V that can be estimated by:

$$v_{m,rs} = \frac{\sum_{i=1}^{n} y_{ri} - f_r(x_i, \hat{\theta})) \times (y_{ri} - f_r(x_i, \hat{\theta}))}{n-1}$$ (21)

For AEs, the sensitivity matrix is $J_r = \partial f_r / \partial \hat{\theta}$, evaluated at n experimental points (sampling times). For DAEs, V can be treated as a sequential experimental design result according to Zullo (1991). With Eq. (20) kept the same, J_r contains the sensitivity coefficient of output y_r with respect to the parameter vector $\hat{\theta}$ evaluated at different sampling times t_s:

$$J_r = \begin{bmatrix} \partial y_r / \partial \theta_1 & \partial y_r / \partial \theta_2 & \cdots & \partial y_r / \partial \theta_m \\ \partial y_r / \partial \theta_1 & \partial y_r / \partial \theta_2 & \cdots & \partial y_r / \partial \theta_m \\ \vdots & \vdots & \ddots & \vdots \\ \partial y_r / \partial \theta_1 & \partial y_r / \partial \theta_2 & \cdots & \partial y_r / \partial \theta_m \end{bmatrix} \begin{matrix} \leftarrow \text{At sampling time } t_1 \\ \leftarrow \text{At sampling time } t_2 \\ \vdots \\ \leftarrow \text{At sampling time } t_n \end{matrix}$$

The diagonal terms of V lead to the following estimation of the confidence region:

$$\sigma_{\hat{\theta}} = F(\alpha, n, m) \times \text{diag}(V^{1/2})$$ (22)

$F(\alpha, n, m)$ represents a probability distribution with confidence level α with n and m degrees of freedom. The smaller σ is the better estimate $\hat{\theta}$ turns out to be. Moreover, σ is closely related to V as in Eq. (22) which paves the way to the following $m \times m$ Fisher information matrix M (where m is the number of model parameters):

$$M(\theta, \varphi) = \sum_{r=1}^{q} \sum_{s=1}^{q} v_{rs}^{-1} J_r J_s$$ (23)

In MBDOE, M helps designing a series of experiments based on the model structure. By carrying out these experiments, the model parameters can be estimated with the best accuracy. The unknown design vector φ contains measured time, initial conditions, control variables, and so on. Minimizing V, corresponds to maximizing the absolute value of M with respect to φ. For a single parameter model, J is $nx1$ and V is a scalar. Parameter estimation and DOE rely on the maximization of $M(\theta,\varphi)$ with respect to θ while and φ correspondingly. The smallest amount of experiments amounts to the best model. It corresponds to the objective function suggested by Espie & Macchietto (1989).

$$F = \max_{\varphi \in \phi} \int_{t_0}^{t_f} \sum_{i=1}^{N_M} \sum_{j=i+1}^{N_M} T_{i,j}(\phi,t)dt \tag{24}$$

where

$$T_{i,j} = \left(f_i(\varphi,\phi_i,t) - f_j(\varphi,\phi_j,t)\right)^T \times \left(f_i(\varphi,\phi_i,t) - f_j(\varphi,\phi_j,t)\right), \text{ and}$$

N_M is the number of candidate model structures. As continuous sampling is not feasible, the integration is replaced by Σ_{tk}. Eq. (24) gives the φ that maximizes the differences among models f_i. Thus, after getting the real experiment profile y_m, the best candidate model predicts y_m most truthfully.

MBDOE has still some drawbacks that require further study:

1. Now and then, it fails to find out the optimal experiment for medium and large scale DAE systems and it generally takes a long time even for small scale systems;
2. There is no trivial/automatic way to classify model parameters sensibly;
3. All criteria depend on optimizing the prediction error variance and V of M in some sense. When M is ill-conditioned, V cannot be numerically calculated, because M cannot be inverted. A possible solution is working with M instead of V;
4. It is difficult to handle models for DAE systems.

5.2 Principal Component Analysis (PCA)

PCA decomposes the data matrix from experiments X by the following expression:

$$X = T^TP + E, \tag{25}$$

according to Coelho et al. (2009), where $X \in \Re^{m \times n}$, with scores $T \in \Re^{n \times n_{pc}}$, loadings $P \in \Re^{m \times n_{pc}}$, residual E and n_{pc} is the number of principal components (PCs). Nice PCA features are:

1. If b_i is the ith eigenvalue of the covariance matrix $(X^T X/n-1)$ in descending order, then the columns t_i of T are orthonormal and explain the relationship between each row:

$$B = \frac{T^TP}{n-1} = diag(b_1, b_2, ..., b_m). \tag{26}$$

2. The columns p_i of P are orthonormal $(I=P^TP)$, and capture the relationship between each column of X. Because $X^T X$ is symmetric, its eigenvalues and eigenvectors are real.

The first few columns of T and P explain most of the variance in X. When $n_{pc}=min(m,n)$, $E=0$. The Cumulative Percent Variance (CPV) is one such method of obtaining the optimal n_{pc} that separates useful information and from noise and the threshold for this method can be set to 90% (Qin & Dunia, 1998; Zhang & Edgar, 2007).

$$CPV(n_{pc}) = \left(\frac{\sum_{j=1}^{n_{pc}} b_j}{\sum_{j=1}^{m} bj} \right) \tag{27}$$

The relationship between PCA and Singular Value Decomposition (SVD) can be explained by the next equations:

$$SVD : \frac{1}{n-1} X^T X = WLC^T \tag{28}$$

$$PCA : \frac{1}{n-1} \left(TP^T\right)^T \times \left(TP^T\right) = \frac{1}{n-1} P \times \left(T^T T\right) \times P^T = PBT^T \tag{29}$$

Since $X^T X$ is a real symmetric matrix, $W(mxm)$ contains the left eigenvectors, $C(mxm)$ has the right eigenvectors and $P=C=W$. The related eigenvalues are in $L(mxm)=B$.

5.3 PCA and Information matrix combined criterion for DOE (P-optimality)

For the sake of simplicity, assume there is only one measured output ($q=1$) and the measurement error is $v_{m,rs}=1$, such that Eq. (20) becomes:

$$V(\theta,\varphi) = [J^T J]^{-1} = M^{-1} \tag{30}$$

The sensitivity matrix J can be viewed as X in the above PCA equations, and M is proportional to V (the scaling factor $(1/n-1)$ in the covariance is contained in $v_{m,rs}$). Assume the eigenvalue and eigenvector matrices of M are Λ and P, respectively. Inserting Eqs. (25), and (29) into Eq. (30) yields:

$$M = J^T J = \left(TP^T\right)^T \times \left(TP^T\right) = P.\Lambda.P^T \text{, and} \tag{31}$$

$$V = M^{-1} = (P.\Lambda.P^T)^{-1} = P^{-T}.\Lambda^{-1}.P^{-1} \tag{32}$$

Since $P^T.P=I=P^{-1}.P$, and $P^T=P^{-1}$, then $V(\theta,\varphi)=P^{-T}.\Lambda^{-1}.P^{-1}=P.\Lambda^{-1}.P^T$. From PCA analysis, V comes from M, by means of SVD or NIPALS. If the smallest eigenvalue in M is λ_m, then λ_m^{-1} will be the largest eigenvalue of V, which indicates the largest variance in the prediction error covariance matrix. The corresponding eigenvector P_m gives the direction of the largest variance in the m parameter space $\Re^m$.

Figure 8 shows the covariance matrix of a two-parameter system. Two eigenvectors p2, p_1 indicate the direction of largest and second largest direction of variance. The projection of long axis (p_2 direction) on θ_1 and short axis (p_1 direction) on θ_2 is proportional to the coincidence region of θ_1 and θ_2, respectively. In Figure 8, when λ_1 is much larger than λ_2, the ellipsoid will degenerate into a line and it is reasonable to look at λ_2 alone. Instead, when θ_2 is well known, one can only focus on shrinking the projection of both ellipsoid axes on θ_1 direction: $\min(|p_1/\lambda_1| \times |p_2/\lambda_2|)$. In order to eliminate the absolute value and take advantage of the unit length of p_i, we use the following expression:

$$Q = \min\left(\frac{p_1(1)^2}{\lambda_1} \times \frac{p_2(1)^2}{\lambda_2}\right) = \max\left(\frac{\lambda_1}{p_1(1)^2}\right) \times \left(\frac{\lambda_2}{p_2(1)^2}\right)$$

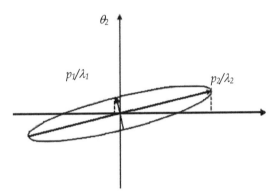

Fig. 8. Geometric interpretation of PCA combined DOE criteria

It is reasonable to reformulate the objective function as follows:

$$F = \min_{\varphi \in \phi} \prod_{i=m-n_{pc}+1}^{m}\left(b_i \sum_j P_{ji}^2\right), \tag{33}$$

where b_i are eigenvalues of V in ascending order ($b_i = 1/\lambda_i$) and λ_i is in descending order) and P is the corresponding eigenvector matrix. The advantage of storing eigenvalues of V in ascending order is that P can be used directly without transformation; otherwise, P for V needs to be transformed by:

$$P_V = P_M = \begin{bmatrix} 0 & \cdots & 1 \\ \vdots & \ddots & \vdots \\ 1 & \cdots & 0 \end{bmatrix}$$

j corresponds to the parameters selected to increase estimation accuracy. To improve the precision of all parameters ($j=1{:}m$) all the PCs are retained and Eq. (33) becomes:

$$F = \min_{\varphi \in \phi} \prod_{i=1}^{m}\left(b_i \sum_j P_{ji}^2\right), \text{ with } \sum_{j=1}^{m} P_{ji}^2 = 1$$

When only the largest eigenvalues of V are used by PCA ($i=m$) and all parameters are to be estimated, Eq. (33) turns out to be

$$F = \min_{\varphi \in \phi} \left(b_m \sum_j P_{jm}^2 \right) \qquad (34)$$

When all the PCs are to be used with specific parameters to be estimated (e.g., the first s), Eq. (33) becomes

$$F = \min_{\varphi \in \phi} \prod_{i=1}^{m} \left(b_i \sum_{j=1}^{s} P_{ji}^2 \right) \qquad (35)$$

After obtaining the eigenvalues of the V matrix (b_i), a series of experiments are designed to minimize b_1, b_2,...,b_m respectively. In general, by minimizing some eigenvalues, the estimation of certain parameters will improve.

When calculating n_{pc}, the eigenvalues of either M or V can be chosen. If V is used, then the last n_{pc} eigenvalues (kept in ascending order such that P does not need to be transformed) and the corresponding eigenvectors should be used to characterize the objective function. When using M, if the first k eigenvalues sum to 90%, then the remaining m-k eigenvalues $n_{pc}=m$-k and eigenvectors are used in Eq. (30). In general, for most model parameters a single eigenvalue cannot comprise information for most parameters (some elements in p_i are close to zero), thus retaining more eigenvalues in the objective function for the first few runs is better. Commonly speaking, the new criterion has the following advantages:

For medium and large-scale DAE systems, it is easier to shrink the scale of the DOE problem by choosing certain parameters out of the entire set to be the focus. By introducing PCA to carry out both eigenvalue calculation and selecting the optimal number of eigenvalues to evaluate, the ill-conditioning of M is avoided. PCA automatically chooses the optimal number of eigenvalues to be investigated, and reduces the problem scale. P gives a clue on grouping the estimated parameters, so it is easy to design an experiment for improving specific parameter estimation, compared with conventional methods.

6. Summary

It is noticed that the factorial design does not determine the optimal values in a single step, but this procedure suitably indicates the path to reach a nice experimental design.

Main effects and the interaction effect are calculated using all the observed responses. Half of the observations belong to one mean, while the remaining half appears in other mean. There is not, therefore, idle information's in the planning. This is an important characteristic of factorial design two-level.

Using factorial design, the calculation of the effects becomes an easy task. The formulation can be extended to any two-level factorial design. The system generated can be solved with the aid of a computer program for solving linear systems.

Modeling focuses on mathematic equations that try to reproduce the real-world behavior accurately over a wide range. Still, regardless of modeling approach chosen, the resulting mathematical models are frequently nonlinear algebraic equations (AE), dynamic algebraic equations (DAE), or partial differential equations (PDE). AE and DAE systems are the most frequently used modeling techniques. Model parameters are in general used to describe special properties such as reaction orders, adsorption kinetics, etc. Hence factorial designs may not be satisfactory for intricate systems.

As a rule, model parameters are not known a priori and have to be estimated from measurements. Moreover, disturbing the system under study very often leads to repetitive measurements and does not produce new data. This leads to the problem of designing experiments prudently to maximize information for specific modeling purposes. An alternative to DOE relying on the previous assumptions is MBDOE.

This work introduces a PCA-based optimal criterion (P-optimal) for model-based DOE that combines PCA with information matrix analysis proposed by Zhang et al. (2007). The main advantages of P-optimal DOE include ease of reducing the scale of optimization problem by choosing parameter subsets to increase estimation accuracy of specific parameters and avoid an ill-conditioned information matrix.

Countless products are produced from the investigation of a large amount of sensors to mine data for analysis. In such cases, the available data maybe correlated, and PCA in addition to other multivariate methods are normally used. PCA is a multivariate technique in which a large number of related variables is transformed into a smaller number of uncorrelated variables (dimensionality reduction).

7. References

Asprey, S.P. & Macchietto, S. (2000) Statistical tools for optimal dynamic model building. Computers and Chemical Engineering, 24:1261-1267.

Asprey, S.P. & Macchietto, S. (2002) Designing robust optimal dynamic experiments. Journal of Process Control, 12:545–556.

ASTM D 638 (2010). Standard Test Method for Tensile Properties of Plastics, Annual book of American Society for Testing of Material (ASTM), U.S.A., Vol. 08.01.

Atkinson, A.C. (2003) Nonconstant variance and the design of experiments for chemical kinetic models. S. P. Asprey and S. Macchietto, editors, Dynamic Model Development, volume 16. Elsevier.

Box, G.E.P. & Draper, N.R. (1987), Empirical model buiding and response surfaces. New York: J. Wiley, 669p.

Carvalho, G.; Silva, M.P.R.; Machado, J.M.P. (2003). Computer Modelling for optimization polimeric blends, Brazilian Meeting SBPMat, 2003, Rio de Janeiro, Brazil.

Coelho, A., Estrela, V. V. & de Assis, J. (2009). Error concealment by means of clustered blockwise PCA. IEEE. Picture Coding Symposium. Chicago, IL, USA.

Espie, D.M. & Macchietto, S. (1989) The optimal design of dynamic experiments. AIChE Journal, 35:223–229.

Franceschini, G.; Macchietto, S. Model-based design of experiments for parameter precision: state of the art. Chem. Eng. Sci. 2008 , 63, 4846 –4872.

Galvanin, F., Macchietto, S. & Bezzo, F. (2007) Model-based design of parallel experiments. Ind. Eng. Chem. Res., 46:871–882.

Galvanin, F., Boschiero, A., Barolo, M. & Bezzo, F. (2011) Model-Based Design of Experiments in the Presence of Continuous Measurement Systems, Industrial & Engineering Chemistry Research, 50 (4), pp. 2167-2175.

Hage, E. & Pessan, L. A., (2001). Improvement in plastics technology. Module 7: polymeric blends. (in Portuguese). ABPol, São Carlos.

Ihm, D.J. & White, J.L., (1996). Interfacial tension of polyethylene/polyethylene terephtalate with various compatibilizing agents. J. of Ap. Polymer Sc., vol. 60, pp. 1-7.

Koning, C.; Duin, M. V.; Pagnoulle, C. & Jerome, R.; (1998). Strategies for Compatibilization of Polymer Blends. Pmg. P&m. Sri., Vol. 23, 707-7.57

Malinov, S. & Sha, W., (2003). Software products for modelling and simulation in materials science. In Proceedings of the Symposium on Software Development for Process and Materials Design, volume 28, Issue 2, pp. 179-198.

Marconcini, J. M. & Ruvolo Filho, A. (2006). Thermodynamic analysis of mechanical behavior in the elastic region of blends of poly (ethylene terephthalate) recycled and recycled polyolefins. (in Portuguese). Polímeros vol.16 no.4 São Carlos Oct./Dec.

Montgomery, D. C. (2005). Design and Analysis of Experiments: Response surface method and designs. New Jersey: John Wiley and Sons, Inc.

Myers, R. H. & Montgomery, D. C. (1995) Response Surface Methodology, Wiley.

Neseli, S.; Yaldiz, S. & Turkes, E. (2011). Optimization of tool geometry parameters for turning operations based on the response surface methodology. .Measurement 44 (2011) 580-87.

Neto, B. B., Scarminio, I. S., & Bruns, R. E., (2003). How do experiments: research and development in science and industry. (in Portuguese). ed. UNICAMP, Campinas, 2 ed.

Pang, Y. X., Jia, D. M., Hu, H. J., Houston, D. J. & Song, M., (2000) Polymer, 41: 357

Paul, D. R. & Bucknall, C. B. (Eds.) (2000). Polymer Blends Volume 1: Formulation, John Willey & Sons, New York.

Qin, S.J. & Dunia, R.H. (1998) Determining the number of principal components for best reconstruction. In IFA C DYCOPS'98, Greece.

Pawlak, A., Morawiec, J., Pazzagli, F., Pracella, M. & Galeski, A. (2002). Recycling of postconsumer poly(ethylene terephthalate) and high density polyethylene by compatibilized blending. J. Appl. Polym. Sci., 86:1,473-1,485.

Rossini, E. L. (2005). Obtenção da blenda polimérica PET/PP/PE/EVA a partir de "garrafas PET" e estudo das modificações provocadas pela radiação ionizante. Tese de Doutorado. Instituto de Pesquisas Energéticas e Nucleares.

Silva, W. S. (2011). Modeling and optimization of ternary of polypropylene (PP), ethylene-propylene-diene monomer (EPDM) and scrap rubber tire (SRT). (in Portuguese). Master's Thesis. Polytechnic Institut of Rio de Janeiro State University.

Tang Q, Lau Yb, Hu S, Yan W, Yang Y, Chen T, (2010). Response surface methodology using Gaussian processes: towards optimizing the trans-stilbene epoxidation over Co2+-NaX catalysts, Chemical Engineering Journal, 156: 423-431.

Walter, E. & Pronzato, L. (1990) Qualitative and quantitative experiment design for phenomenological models - a survey. Automatica, 26:195−213.

Wessler, Katiusca. (2007). Systems of P (3HB) and P (3HB-co-3HV) with Poly-triol: Phase Behavior, Rheology, Mechanical Properties and Processability. Sistemas de P(3HB) e P(3HB-co-3HV) com Policaprolactona-Triol (in Portuguese), Dissertation. Center of Technological Sciences. State University of Santa Catarina.

Zhang, Y.; Edgar, T. F. (2008) PCA combined model-based design of experiments (DOE) criteria for di fferential and algebraic system parameter identification. Ind. Eng. Chem. Res. 2008 , 47, 7772 -7783.

Zhang, Y. & Edgar, T.F. (2007) Online batch process monitoring using modified dynamic batch PCA. ACC, pages 2551-2556, NY, IEEE.

Zullo, L. (1991) Computer aided design of experiments. An engineering approach, PhD thesis, University of London.

3

Applications of Principal Component Analysis (PCA) in Materials Science

Prathamesh M. Shenai[1], Zhiping Xu[2] and Yang Zhao[1]
[1]Nanyang Technological University
[2]Tsinghua University
[1]Singapore
[2]China

1. Introduction

Nowadays we are living in the information age with the fast development of computational technologies and modern facilities. Larger data sets are produced by experiments and computer simulations. In contrast to conventional scientific approaches where simple models are built to fit the data, automated procedures are urged to obtain insights into the core messages carried by the large volume of data.

Many problems encountered in materials science involve complicated data models. For example, in biological materials, the collective motion of protein domains usually defines the structural and biological activity of proteins, which should be separated from the irrelevant localized motion of atoms and molecules with high-frequencies. An efficient approach to capture the essential subspace of protein dynamics can remarkably reduce the complexity and directly uncovers the underlying physics (Amadei et al., 1993). On the other hand, nanostructures, which are widely used in nanoscale devices, also have several functional modes that are closely tied to their operation. To visualize them in a thermal and noisy environment requires some insightful treatment (Xu et al., 2008).

Principal component analysis (PCA), as invented by Karl Pearson in 1901, is a procedure to convert a set of correlated variables into uncorrelated ones called principal components (Joliff, 2002). Using mathematical algorithms such as eigenvalue decomposition of the covariance tensor or single value decomposition (SVD), PCA methods find successful applications in many fields as covered in this book. Figure 1 shows the principal modes of ubiquitin in solvent and carbon nanotubes (CNTs) under water flow, as mined from their correlated dynamics in solvents.

In this chapter we will introduce the applications of PCA method in materials science, which not only assist to find useful patterns from the detailed dynamics of atoms and molecules, but also advances the development of PCA technique itself.

2. The mathematics and algorithms of PCA

There are many areas of scientific explorations that lead to enormous quantities of data. Post-processing of such a huge data to extract only the most valuable information is often a

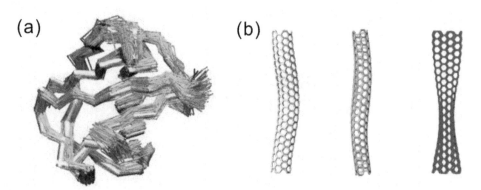

Fig. 1. Applications of principal component analysis (PCA) methods in (a) protein dynamics (Yang et al., 2009) and (b) dynamics of carbon nanotubes under water flow (Chen & Xu, 2011).

tedious task. In a very broad perspective, PCA belongs to a particular set of techniques aimed at reducing a large dataset to a smaller one which can describe the essential characteristics of the underlying system at hand. Molecular dynamics (MD) is a powerful and widely utilized approach in simulating various materials properties and in this chapter, we will focus on the usefulness of PCA in analyzing trajectories generated by MD.

2.1 PCA on MD trajectories

A typical MD trajectory consists of the information of time-evolution of the coordinates of all the constituent atoms forming the system being studied. Commonly used MD timesteps are on the order of 1 fs while the simulation time may range from a few to tens of nanoseconds, in any moderately sized configuration. A single resultant trajectory can thus easily contain a huge amount of data. For an N-atom system, the input dataset for PCA can be constructed as a trajectory matrix in which each column contains a cartesian coordinate for a given atom at each output timestep ($x(t)$). Prior to performing PCA, it is ususally necessary to remove any net translational or rotational motion of the system by fitting the coordinate data to a reference structure to obtain the proper trajectory matrix (X). The standardized trajectory data is then utilized to generate a covariance matrix (C), elements of which are defined as

$$C_{ij} = \langle (x_i - \langle x_i \rangle)(x_j - \langle x_j \rangle) \rangle \tag{1}$$

where $\langle ... \rangle$ denotes an average performed over the all the timesteps of the trajectory. The next step consists of diagonalization of the symmetric $3N \times 3N$ covariance matrix and can be achieved via eigenvector decomposition method as

$$C = T \Lambda T^{\mathrm{T}} \tag{2}$$

where T is a matrix of column eigenvectors and Λ is a diagonal matrix containing the corresponding eigenvalues. This procedure thus transforms the original trajectory matrix in a new orthornormal basis set composed of the eigenvectors. The eigenvalues themselves are indicative of the mean squared displacements of atoms along the corresponding eigenvector. There will be $3N$ resulting eigenvalues if the number of configurations (M) is greater than $3N$. If $M<3N$ there will be the number of eigenvalues will be reduced to M.

The simplest manner of visualizing these results requires sorting the eigenvectors in a descending order in their eigenvalues. The plot of eignevalues against the index of the corresponding eigenvector can then be obtained and is called a 'scree plot'. Characteristically, a scree plot shows that only a few first eigenvectors possess large eignevalues with the higher indexed vectors having eignevalues many orders of magnitude smaller. As a result, most of the variance in the original data is contained and described by only a few first modes. It is then imperative to presume that the motions along these 'essential eigenmodes' dominate the dynamics of the systems and contain the most important global information.

In simple systems, visualization of the components of individual eigenvector can be helpful to gauge the nature of the eigenmode. Followed by identification of a subset of important eigenmodes, further analysis detailing each mode can be undertaken by projecting the original trajectory along a given (or a set of) eigenvector. The corresponding projection matrix (P) can be obtained as

$$P = XT. \tag{3}$$

The time evolution given by the projection matrix yields a manner in which the excitation amplitude of a given eigenvector can be examined. The column vectors in P (p(t)) are called as the 'principal components'.

To analyze the motion along any given eigenvector, the column vector from P multiplied by the corresponding eignevector in T^T yields a reduced trajectory containing motion only along the selected mode. Such filtering of modes can be performed for a single or more than one eigenmodes as well and the resulting trajectory provides a visual guidance to the nature of the mode.

A quantitive measure of similarity (S) between different principal modes can be obtained by taking inner product of the corresponding eigenvectors (v_i amd w_j) from T as follows:

$$S_{ij} = v_i \cdot w_j \tag{4}$$

The same concept can be further extended to calculate a measure of overlap ($O(v,w)$) between an essential subspace spanned by eigenvectors w_j (j=1,2,..,n) and another spanned by eigenvectors v_i (i=1,2,..,m) as (Amadei et al. 1999; Hess 2002):

$$O(v,w) = \frac{1}{n}\sum_{i=1}^{n}\sum_{j=1}^{m} S_{ij} \tag{5}$$

The overlap will be equal to unity if the subspace spanned by v_i is a subset of w_j.

2.2 Computational implementations

Apart from long-time MD simulations to generate sufficient trajectory data, the diagonalization of the $3N$ X $3N$ covariance matrix poses the most computationally exhaustive step during PCA. The computational expense as well as memory requirements increase roughly with the square of the number of atoms in the system. As a result, for quite large systems (which can easily be the case when considering large biomolecules), use of efficient algorithms such as QR decomposition is required for matrix diagonalization. Due to the widespread use of PCA, some existing molecular dynamics programs including open source packages such as GROMACS (Hess et al., 2004) and AMBER (Case et al., 2005) and commercially available Accelrys Materials Studio have incorporated implementations of PCA. Another helpful utility is Interactive Essential Dynamics (IED) which can use the output of PCA performed with GROMACS/AMBER to visualize filtered trajectories via a graphical user interface (Morgan, 2004).

2.3 Demonstrative calculations on a single walled carbon nanotube

Emergence of CNTs and graphene as potential candidates for nanoscale machines has led to their exhaustive probing by using molecular dynamics. It is likely that PCA can prove extremely useful in uncovering many novel dynamical features in such scenarios. In this section, we thus apply PCA to MD simulations of a single walled carbon nanotube (SWNT) with its chirality specified as (5,5). Two different approaches viz. fine-grained and coarse-grained models are studied. The fine-grained approach consists of the regular full atomistic simulations on the SWNT configuration. The other approach adopted from Buehler et al. consists of approximating the structure of the SWNT as finite-sized beads connected with stiff springs (Buehler, 2006).

2.3.1 Fine-grained (fully atomistic) approach

A long (5,5) SWNT configuration with lengths ~ 100 nm (8000 atoms) is considered, a schematic of which is shown in figure 2(a). The intratube C-C interactions are described by Adaptive Intermolecular Reactive Bond Order (AIREBO) potential (Stuart et al., 2000) and MD simulations are performed on the equilibrated structures in a canonical ensemble at 300 K. Temperature control is exercised through the use of Berendsen thermostat (Berendsen et al., 1984).

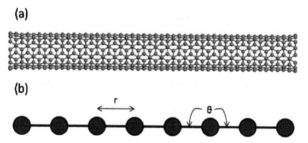

Fig. 2. (a) A schematic of atomistic model of a (5,5) SWNT and (b) a corresponding coarse-grained bead-spring model.

All the simulations are performed using the massively parallelized open source MD software LAMMPS (http://www.cs.sandia.gov/~sjplimp/lammps.html) with a timestep of 1 fs (Plimpton, 1995). At first, the system is thermalized at 300 K for 100 ps. The production run is carried out for 10 ns and the obtained trajectories are subjected to PCA using various tools available in GROMACS. For analyzing the long tube, the production run trajectory is sampled every 50 ps. This sampling rate is chosen to focus on low frequency bending modes and to match the time-scale for a fair comparison with coarse-grained model described in the next subsection.

2.3.2 Coarse-grained approach

Fully atomistic simulations become increasingly computationally prohibitive as the number of atoms in the system grows. As a result, especially when the study of the structural properties at micro-scale is required, the precise atomistic information is rendered redundant. A coarse-graining approach delineated in the work of Buehler et al. can be useful to circumvent the computational expenses and allow for investigation on longer scales. In this approach, a SWNT is essentially modeled as a linear chain of beads connected via springs as depicted in figure 2(b). The properties of individual beads (such as mass) and the springs (such as tensile stretch, angle bend and torsion) can be determined from full molecular dynamics. In this work, we adopt the same approach and coarse-grain a 100 nm long (5,5) CNT as a 100 beads-chain with an equilibrium inter-bead separation distance of 1 nm. All the required parameters can be found in (Buehler 2006) and the dynamics of this system is simulated using LAMMPS. The time-step chosen for the dynamics is 50 fs and the production run is carried out for 10 ns. Using a sampling rate of every 1000 timesteps, PCA is performed on the coordinates' data of all the beads in an analogous way as explained before.

2.3.3 PCA results

Figure 3 shows the scree plot for both the coarse-grained and the atomistic model of the CNT for the first 30 modes. It can be observed in either case that only a few of the first modes occupy high eigenvalue position and thus contain the essential information of the bead dynamics. Modes at higher indices correspond to smaller eigenvalues. Although both the models show a high-eigenvalue first mode, as compared to the coarse-grained model the atomistic model shows a more gradual decrease in the eigenvalues. One of the reasons in the difference is that the correspondance between the similarly indexed modes from the two models is not strictly perfect and as described later, the atomistic system displays a slighly more complicated hierarchy of modes.

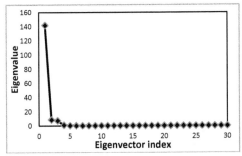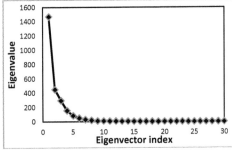

Fig. 3. Eigenvalues against the index of eigenvector obtained from PCA on (a) coarse-grained model of (5,5) SWNT and (b) full atomistic model.

As the coarse-grained system is quite simple, it is intuitive to take a look at the components of individual eigenvectors. For the first five modes, the eigenvector components corresponding to x, y and z coordinates of each bead's center are shown in figure 4. It becomes apparent that the eigenvectors indeed represent the fundamental vibrational modes of the bead-spring system and its harmonics which resemble to that of the vibrational modes of a strectched spring. Being a slightly more complex system than a vibrating string, the individual principal modes in the bead-spring model can further be seen as the superimposed vibrational modes along X and Y directions. A rough estimation of the mode frequency can also be obtained from their projections on the original trajectory. Note that within a coarse-grained model, any of the modes constituting radial displacements cannot be present and thus, the top principal modes revealed are the bending-like oscillatory modes.

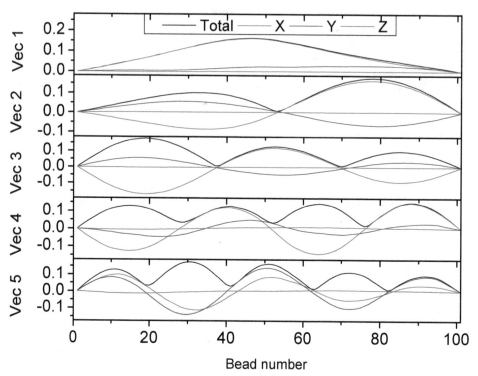

Fig. 4. The components of the most significant five eigenmodes for the (5,5) SWNT. The principal modes resemble fundamental vibrational mode and its harmonics for a stretched string.

The components of the first five eigenvectors of the atomistic model can also be examined in a similar manner and are shown in figure 5 such that the atom numbers are indexed consecutively along the circumference and length. With respect to full atomic contributions (all the X, Y and Z variables) certain qualitative similarities between the mode patterns among the two models can be easily observed. Similar to the bead-spring model, the first eigenmode is the fundamental bending mode of the CNT while the next higher modes represent more or less the sequentially higher harmonics as well. However, a closer look reveals a slightly more complexity, e.g. in the nature of 3rd and 4th principal modes. It can be noted that unlike the bead-spring model, the third mode here appears to be a superimposition of the third harmonic along X-axis and second harmonic along the Y –axis of CNT.

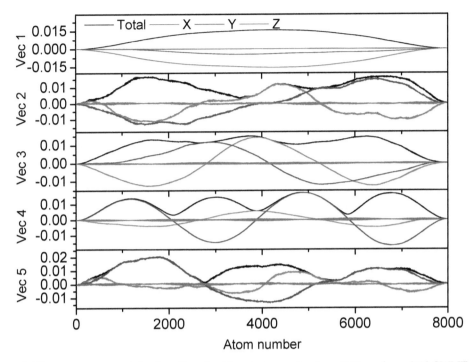

Fig. 5. The components of the first five significant eigemodes for a 100 nm long (5,5) SWNT.

These simple representative calculations thus demonstrate how PCA can help identify various essential modes in a molecular system. In addition to MD simulations, mesoscale simulations of CNTs have started to appear in recent literature. Here, we have purposefully chosen comparisons of atomistic simulations with coarse-grained model of a long carbon nanotube to focus mainly on the out-of plane bending modes. Comparisons between different simulation models for studying material properties at different scales can thus be seen greatly assisted with the use of PCA.

3. Applications in biology, advantages and limitations of PCA

3.1 PCA in biomolecular MD

Biological systems are of immense research interest not only because of the fundamental mysteries of living systems involved, but also because of the possibilities of imitating principles of natural designs in advanced technologies. Proteins, which are the basic building blocks of life, exhibit a striking functional dependence on their conformation. At the cellular level, a variety of biological machinery work precisely amidst extremely noisy environments. Extraction of physical principles that govern such directional dynamics may prove crucial in constructing their artifical counterparts at the molecular level. MD simulations of large biomolecules in fact presented the need of introduction of PCA (Amadei et al., 1993). Amadei et al. proposed that except for the degrees of freedom that belong to the 'essential subspace' of proteins, all the other modes are largely irrelevant

Gaussian fluctuations. The technique quickly became popular in analyzing MD simulations of a large number of biomolecules.

Folding of a protein in a well defined characteristic three-dimensional structure from a random coil structure is one of the most crictical biophysical processes, and PCA has proved vastly useful in its exploration using MD. Ligand binding in proteins such as Myoglobin is strongly influenced by very specific conformational changes near the binding sites. Touriner and Smith investigated MD simulations of hydrated myglobin and found a single principal mode primarily responsible for a dynamical transition appearing at about 180 K (Tournier & Smith, 2003). A class of proteins called membrane proteins such as gramicidin, serves a crucial role of formation of selective ion channels that regulate fluxes across the cell membrane. Recently, Kurylowicz et al. probed the role of anharmonic principal modes such as tilting of peptide planes towards its function as ion channel in gramicidin-A (Kurylowicz et al., 2010). Here we resctrict ourselves to enlist only a few representative examples from the vast literature exists pertaining to PCA in biololecules.

3.2 Inadequacies of PCA

PCA can be performed on either a long single MD trajectory or an ensemble of short trajectories. The latter route is usually advocated since in biomolecular MD simulations, since it is well known that PCA presents difficulties with respect to proper sampling (Balsera et al. 1996; Caves et al. 1998). An excellent analysis about reliability of PCA with respect to sampling issues can be found in the work of Skjaerven et al. (Skjaerven et al., 2011). PCA perfomed on multiple independent runs of the protein systems under the same simulation conditions except the initial atomic velocities, revals noticable differences (de Groot et al., 1998; Skjaerven et al., 2011). While PCA on a single trajectory unambiguosly identifies essential modes during the simulation time, significant differences that can be found among independent runs suggest inadequancy of the sampling of dyamics in the trajectory.

Computational limitations make MD possible on ns timescale while many conformational transitions in proteins, nucleic acids may occur on ms or greater scale. As a result, a single MD trajectory may not entail all possible modes that are essential towards dynamical conformational changes. A direct consequence of such considerations is that even for a single trajectory, the principal modes obtained during one observation window may differ from an another window. While this remains true, a very long (few hundreds of ns) MD simulation may not necessarily yield highly convergent eigenvectors from PCA as compared to simulations with a timespan on the order of tens of ns. Even though efforts have been put in devising methods of enhanced sampling of essential dynamics (Amadei et al., 1999; Hess, 2002), convergence of eigenmodes remains a critical issue.

3.3 Variants of standard PCA

In certain cases, it is also possible and perhaps more useful to disregard less important internal coordinates such as bond lengths and restrict the consideration to dihedral angles. Implementation of PCA based on dihedral angles is commonly referred to as dihedral PCA (dPCA) and was introduced by Mu et al. (Mu et al., 2005). The developement of this approach is mainly aimed at reduction in the dimensionality of the input covariance matrix itself. It has been shown that the dPCA yields results generally equivalent to those obtained with the conventional cartesian PCA (Altis et al., 2007). Further, instead of MD simulations, it is also possible to use the experimentally generated structural data such as from Nuclear Magnetic Resonance (NMR) or X-ray techniques, for performing PCA. In such a case, an ensemble containing a sufficient number of structural models of the biomolecule needs to be determined from the aforementioned experiments (Howe, 2001; Yang et al., 2009). Although analysis on structural analysis cannot resolve precise atomic motion and is thus of 'coarse' nature as compared to MD simulations, it can still provide a crude approach to compare MD models with experimental data (van Aalten et al., 1998).

3.4 Comparison with Normal Mode Analysis (NMA)

In its standard form, NMA is essentially a harmonic analysis technique which relies on an assertion that the functionally important modes can be extracted as the low frequency normal vibrational modes. The underying assumption is that the conformational energy surface for a given system is approximately parabolic at the global energy minimum. NMA has also been vastly used in structural biology to gain an understanding of the fundamental functional modes in macromolecules such as proteins, lipids and nucleic acids. For a comprehensive account of NMA in reference to biological simulations, reader is referred to an excellent review by Hayward and Go (Hayward & Go 1995).

As NMA demands the structure to be in its lowest energy state, it first needs to be subjected to thorough energy minimization. The next step consists of evaluation of the 'Hessian' (H), which is a matrix of second derivatives of the energy (U) with respect to displacements along cartesian coordinates (x_i), and is calculated as

$$H_{ij} = \frac{\partial U}{\partial x_i \partial x_j} \tag{6}$$

The diagonalization of the mass weighted Hessian ($M^{-1/2}HM^{-1/2}$) where the diagonal matrix M contains the information of atomic masses then yields the eigenvectors and corresponding eigenfrequencies. As opposed to PCA, the normal modes are sorted in ascending manner according to their frequencies.

The fundamental difference between NMA and PCA is in the harmonicity of the resulting modes. Due to the underlying assumption, NMA invariably is restricted to small amplitude harmonic fluctuations around the energy minimum. PCA on the other hand, deals with the positional fluctuations and is thus well suited to study anharmonic vibrations. Furthermore, existing evidence suggests the functional modes in biomolecules to be anharmonic in nature (Amadei et al., 1993; Hayward et al., 1995), which implies that at the physiological temperatures, the underlying assumption of NMA becomes too drastic to be relevant. As a result PCA can be viewed as the more apt technique among the two for exploring dynamical transitions. Yet, standard NMA and its variants such as elastic network NMA have been quite extensively utilized in understanding low frequency functional modes in proteins. Due to the need of long-time MD trajectories, PCA is much more computationaly exhaustive as compared to NMA whereas NMA simply requires a single lowest energy configuration.

4. Applications of PCA in nanomaterials

Compared to biological systems, application of PCA in materials simulations has been sparse. The possible reasons include a lack of material systems for which detailed molecular motions influence the macroscopic dynamical behavior significantly. However, past couple of decades have witnessed a huge increase in interest in nanoscale transport and mechanical phenomena such as carbon nanotube based hyper-GHz mechanical oscillators, resonators, rotational bearings and actuators (Bourlon et al., 2004; Cumings & Zettl 2000). Double walled CNTs (DWNTs) and multiwalled CNTs are blessed with a rare combination of strong mechanical elements in the form of constituent SWNTs which interact weakly via van der Waal's forces. This sets up an ideal scenario to construct devices in which relative inter-tube rotation or translation can be achieved at the expense of negligible frictional loss. While the required technology at the atomic level is yet to mature, theoretical and molecular dynamics based approaches have opened up proactive paths of investigating characteristics of such nanomachines. This is one of the promising fields of materials research that PCA can fruitfully debut in.

4.1 Analyzing dynamics of CNT based nanomachines

In our previous work, we were able to deduce analogies between dynamics of a rapidly translating SWNT inside a larger SWNT and an aircraft flying near supersonic speed (Xu et al., 2008). It was discovered that for most of the travelling speeds, the core tube can translate without any significant frictional dissipation. However, at certain specific values of axial velocities, abrupt increase in frictional effects can take place. Such kind of energy dissipation points to possibility of resonance effects at particular travelling velocities and is in contrast with the phononic friction commonly observed in nanodevices. We used PCA to gain insights into the nature of modes present in the nanotube-shuttle systems, in one of the first direct applications of PCA in analysing nanodevices.

Using detailed PCA, the underlying principal modes constituting the total motion in the MD trajectories were identified as shown in figure 6. The striking feature in the scree plots corresponding to those initial velocities (1000 m/s and 1900 m/s) at which frictional enhancements appear can be observed in the excitation of high indexed vibrational modes. It was found that at the detrimental critical speed range, a resonance occurs between the 'washboard frequency' and the radial breathing mode (RBM) frequency of the constituent DWNT. The coupling of RBMs with other non-rigid body modes such as bending modes

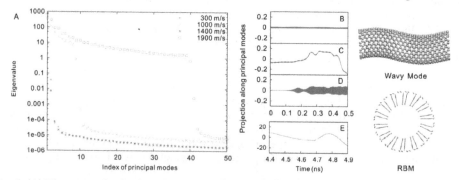

Fig. 6. (A) The scree plot for (7,7)/(12,12) DWNT configuration with different initial travelling speeds of inner nanotube, of which 1000 m/s and 1900 m/s lead to resonant frictional effects. (B)-(D) The projections of wavy, RBM and bending modes for 1000 m/s. (E) Projection of RBM-like mode towards the end of simulation (Xu et al., 2008).

further ensures a nonreversible energy dissipation. Resonant excitation of RBM is evident in figure 6(C) and figure 6(E) that promotes the excitation of various non-rigid body modes such as wavy or bending modes (see figure 6B and 6D). As a result, uncovering new resonant frictional regimes in nanoscale devices by using PCA was demonstrated successfully.

In the case of rotational nanobearings based on DWNTs, Shenai et al. found operational behavior for short sleeved configuration reminiscent of the trans-phonon effect in the translational counterparts (Shenai et al., 2010). It was found that the rotational bearing exhibits a step-like dissipative operation in which at certain angular speeds, the bearing appears to rotate in nearly frictionless manner. The stable rotation, can get hampered however, in such a manner that the angular velocity dissipates more or less abruptly until the bearing stabilizes in a next lower favorable angular speed range. In this case as well, by application of PCA to the bearing trajectory during stable operation and during dissipative operation, it was detected that excitation of dissipative wavy modes takes place during the decay period shown as the 4th eigenmode in figure 7(d).

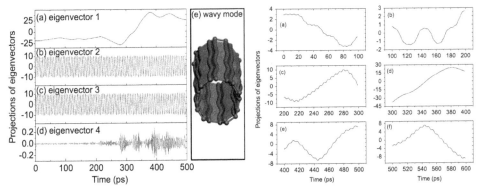

Fig. 7. (a)-(d) Projections of first four eigenvectors on the trajectory of a DWNT based rotational nanobearing. While axial translation reveals itself as the first eigenmode, the rotation is represented by 2nd and 3rd modes together. (e) Depiction of the dissipative wavy mode as the 4th eigenmode. Right panel shows similar analysis for the first eigenvector in different time periods from the initial 600 ps (Shenai et al., 2010).

As a rotational nanobearing was studied in this work, the three lowest eigenmodes are the rigid body modes such as translation (1st eigenvector) and rotation of sleeve (2nd and 3rd eigenvector in combination). More interestingly, it was found that the leakage of the rotational kinetic energy of the sleeve to dissipative wavy modes occurs via another channeling mode – the axial translation of the sleeve. Due to the atomic arrangement of a DWNT, the interaction energy surface between the two tubes exhibits periodic corrugation with respect to relaive axial displacement. Due to typically small corrugation against axial sliding and the small mass of the sleeve, excitation of such translational mode can take place through extraction of a small part of the rotational kinetic energy. When the energy occupation of the axial sliding mode is low, the motion occurs in step-like manner between adjacent energy wells. However when it acquires highly enough translational energies, its enhanced coupling with the higher indexed wavy modes leads to chanelling of the excess energy as shown in the right panel of figure 7. As soon as the axial oscillation dies, the undesirable channel to the wavy modes gets closed as well, thereby suppressing further decay in rotational kinetic energy. The intricacies of the axial sliding motion and its role in the corresponding excitation of wavy modes was thus successfully resolved.

Negi et al. performed a rigorous study of normal modes via singular value decomposition (SVD) to analyze MD trajectories of single walled carbon nanotubes (Negi & Chaturvedi, 2010) under NVE and NPT conditions. Their approach essentially produces results similar to those with the standard PCA. The full spectrum of principal modes including RBMs and other non-rigid body modes was successfully extracted, see for example, figure 8. In the detailed analysis, they categorized the principal modes according to the uniformities in the displacement characteristics of tube atoms along radial, axial and angular directions. In another subsequent study involving rotational nanomotors driven by external electric field, similar SVD analysis was put to use in understanding the operational regimes and characteristics (Negi et al., 2010).

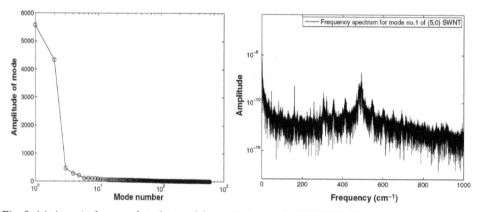

Fig. 8. (a) A typical scree plot obtained from PCA on a (5,0) SWNT. (b) Corresponding power spectrum showing a peak corresponding to frquency of RBM, which is first principal mode (Negi & Chaturvedi 2010).

The aforementioned studies involving the most basic types of CNT based nanomachines demonstrate the usefulness of PCA in analyzing their dynamical features. In future studies probing frictional dissipation or energy channeling between different modes in various nanomechanical devices, PCA can be expected to prove significantly helpful in providing valuable insights.

4.2 Applications in non-linear dynamics of other materials

In another interesting approach by Battisti et al., PCA was innovatively used to study coherent and chaotic dynamics of a small molecule, butane (Battisti et al., 2009). Characterization of chaoticity in the butane molecule was based on evaluation of Lyapunov exponent (LE), which essentially determines the exponential rate of divergence between two trajectories in the phase space separated by a very small distance in their initial conditions.

In this study, a conjecture that in chaotic systems at low energies, different degrees of freedom may entail different degrees of chaoticity was examined. The 'essential' degrees of freedom were obtained as the eigenmodes obtained from PCA on the MD trajectories of butane. Using the individual trajectories reconstructed by projecting different eigenvectors on the original trajectory, it was possible to calculate LE for individual degree of freedom (in terms of principal modes). It was revealed that depending upon the system temperature, there exists a hierarchy of degrees of freedom with respect to 'coherence time' – a measure of degree of order. Certain degrees of freedom exhibit more chaoticity than the system as whole, may exhibit lower chaoticity as shown in figure 9.

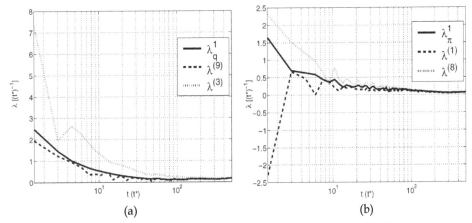

Fig. 9. Differential evolution of Lyapunov exponent during short time for the trajectories filtered along different principal modes for a butane molecule investigated with molecular dynamics in (a) coordinate space at 180 K, and (b) velocity space at 147 K (Battisti et al., 2009).

In general, at relatively high temperatures, the first two eigenmodes turn out to be the most coherent degrees of freedom. Such hierarchy was further shown to vary significantly at varying temperatures as well as under the particular subspaces (coordinate or velocity) at which calculations are performed.

5. Conclusions and perspectives

In this chapter, we have presented a comprehensive account of PCA and its applications in different fields of materials science, in particular. An overview of the underlying theory is presented followed by demonstration of its applications in the study of SWNTs, based on two different approaches – coarse-grained simulations and fully atomistic fine-grained simulations. The results emphasizing the importance of 'essential subspace' and identification of lowest principal modes are presented with respect to the two models along with comparisons between them.

While it has been extensively applied in the studies of biomolecules over the past two decades, possibilities of its usage in the study of materials, have started to emerge only recently. The vast applications of PCA in structural biology have led to developements of its variants such as coarse-grained PCA or dihedral-PCA. Despite being highly successful, concerns regarding accuracy and robustness of PCA such as sampling issues must be addressed very carefully. Such limitations have not been thoroughly investigated when PCA is employed in the study of materials. Strictly speaking, direct application of PCA in core materials science is sill quite limited. Yet, the emerging field of nanomachines based on carbon nanotubes or graphene focussed on MD, undoubtedly stands out as the most promising area where PCA appears to be quite useful in understanding dynamical characteristics.

6. References

Altis, A.; Nguyen, P. H.; Hegger, R. & Stock, G. (2007). Dihedral angle principal component analysis of molecular dynamics simulations. *Journal of chemical physics*, Vol. 126, pp. 244111

Amadei, A.; Linssen, A. B. M. & Berendsen, H. J. C. (1993). Essential dynamics of proteins. *Proteins: Structure, Function, and Bioinformatics*, Vol.17, No.4., pp. 412-425

Amadei, A.; Ceruso, M. A. & Nola, A. D (1999). On the convergence of conformational coordinates basis set obtained by the essential dynamics analysis of proteins' molecular dynamics simulations. *Proteins*, Vol.36, pp. 419-24

Balsera, M. A.; Wriggers, W.; Oono Y. & Schulten, K. (1996). Principal Components Analysis and Long Time Protein Dynamics. *Journal of Physical Chemistry.* Vol.100, pp. 2567-72

Battisti, A.; Lalopa, R. G.; Tenenbaum A. & and D'Alessandro M. (2009). Ordered and chaotic dynamics of collective variables in a butane molecule. *Phys. Rev. E*, Vol.79, pp.46206

Berendsen H. J. C.; Postma J. P. M.; van Gunsteren W. F.; DiNola A. & Haak J. R. (1984). Molecular dynamics with coupling to an external bath. *J. Chem. Phys.* Vol.81, pp. 3684-90

Bourlon B.; Glattli, D. C.; Miko, C.; Forro, L. & Bachtold A. (2004). Carbon Nanotube Based Bearing for Rotational Motions. *Nano Letters*, Vol.4, No.4, pp. 709-712

Buehler, M. J. (2006). Mesoscale modeling of mechanics of carbon nanotubes: self-assembly, self-folding, and fracture. *J. Mater. Res.*, Vol.21, pp. 2855-69

Case, D. A.; Cheatham III, T. E.; Darden, T.; Gohlke, H.; Luo, R.; Merz Jr., K.M. ; Onufriev, A.; Simmerling, C.; Wang, B. & Woods R. (2005). The Amber biomolecular simulation programs. *J. Computat. Chem.*, Vol.26, pp. 1668-88.

Caves, L. S. D.; Evanseck, J. D. & Carplus, M. (1998). Locally accessible conformations of proteins: Multiple molecular dynamics simulations of crambin. *Protein Science*, Vol.7, No.3, pp. 649-66

Chen, C.; Ma, M.; Jin, K.; Liu, J. Z.; Shen, L. M.; Zheng, Q. S. & Xu, Z. (2011) Nanoscale fluid-structure interaction: Flow resistance and energy transfer between water and carbon nanotubes. *Physical Review E*, Vol. 84, pp. 046314

Cumings J. & A. Zettl (2000). Low-friction nanoscale linear bearing realized from multiwall carbon nanotubes. *Science,* Vol.289, No.5479, pp. 602-604 de Groot, B.; Hayward, S.; van Aalten, D.; Amadei, A. & Berendsen, H. (1998). Domain motions in bacteriophage T4 lysozyme: A comparison between molecular dynamics and crystallographic data. *Proteins*, Vol.31, pp. 116-27

Hayward S. & Go, N. (1995). Collective variable description of native protein dynamics. *Annu. Rev. Phys. Chem.*, Vol.46, pp.223-50

Hayward S.; Kitao, A.; & Go, N. (1995). Harmonicity and anharmonicity in protein dynamics: a normal modes and principal components analysis. *Proteins: Structure, Function, and Bioinformatics*, Vol.23, No.2, pp. 177-86

Hess B. (2002). Convergence of sampling in protein simulations. *Phys. Rev. E*, Vol.65, pp. 031910

Hess, B., Kutzner, C.; van der Spoel, D. & Lindahl, E. (2008). GROMACS 4: Algorithms for Highly Efficient, Load-Balanced, and Scalable Molecular Simulation. *J. Chem. Theory Comput.*, Vol.4, pp. 435-47

Howe ,P. W. (2001). Principal components analysis of protein structure ensembles calculated using NMR data. *J. Biomol NMR* Vol.20, No.1, pp. 60-70

Jolliffe, I. T. (2002). *Principal Component Analysis*, Springer

Kurylowicz, M.; Yu, C. H. & Pomes, R. (2010). Systematic Study of Anharmonic Features in a Principal Component Analysis of Gramicidin A. *Biophys. J.* Vol.98, No.3, pp. 386-95

Morgan, J (2004). Ineractive essential dynamics. *J. Computer-Aided Molecular Design*, Vol.18, pp. 433-36

Mu, Y.; Nguyen, P. H. & Stock, G. (2005). Energy Landscape of a Small Peptide Revealed by Dihedral Angle Principal Component Analysis. *Proteins*, Vol.58, pp.45

Negi, S. & Chaturvedi, S. (2010). Normal mode analysis of a single-walled carbon nanotube based on molecular dynamic: A singlular value decompopsition study. *Int. J. Nanosci.*, Vol.9, No.5, pp. 471-86

Negi, S.; Warrier, M. & Chaturvedi, S. (2010). Determination of useful parameter space for a double-walled carbon nanotube based motor subjected to a sinusoidally varying electric field. *Comp. Mater. Sci.*, Vol.50, pp. 761-70.

Plimpton, S. (1995). Fast parallel algorithms for short-range molecular dynamics. *J. Comp. Phys.*, Vol.117, pp. 1-19.

Skjaerven, L.; Martinez, A. & Reuter, N. (2011). Principal component and normal mode analysis of proteins; a quantitative comparison using the GroEL subunit. *Proteins: Structure, Function, and Bioinformatics*, Vol.79, No.1, pp. 232-243

Shenai P. M.; Ye, J. & Zhao, Y. (2010). Sustained smooth dynamics in short sleeved double-walled carbon nanotubes. *Nanotechnology*, Vol.21, No.49, pp. 495303

Stuart S. J.; Tutein A. B. & Harrison J. A. (2000). A reactive potential for hydrocarbons with intermolecular interactions . *J. Chem. Phys.*, Vol.112, pp. 6472-86

Tournier, A. L. and Smith, J. C. (2003). Principal components of the protein dynamical transition. *Phys. Rev. Lett.*, Vol. 91, No.20, pp. 208106

van Aalten, D. M. F.; Grotewold, E. & Joshua-Tor, L. (1998). Essential dynamics from NMR clusters: Dynamic properties of the Myb DNA-binding domain and a hinge-bending enhancing variant. *Methods-A Companion to Methods in Enzymology*, Vol.14, No.3, pp 318-28

Xu, Z.; Zheng, Q.; Jiang, Q.; Ma, C.-C.; Zhao, Y.; Chen, G.; Gao, H. & Ren, G. (2008). Trans-
    phonon effects in ultrafast nano-devices. *Nanotechnology*, Vol.19, No.25, pp. 255705-
    255705

Yang, L.-W.; Eyal, E.; Bahar, I. & Kitao, A. (2009). Principal component analysis of native
    ensembles of biomolecular structures (PCA_NEST): insights into functional
    dynamics. *Bioinformatics*, Vol.25, No.5, pp. 606-614

4

Application of Principal Component Analysis in Surface Water Quality Monitoring

Yared Kassahun Kebede and Tesfu Kebedee
Ethiopian Institute of Agricultural Research
Ethiopia

1. Introduction

Surface water systems such as rivers, lakes and ground water are affected by the natural processes such as erosion of minerals and dissolution of nutrients from the overlying rocks as well as anthropogenic influences from urban, industrial and agricultural activities. These degradation of surface water quality resulted in altered species composition and decreased overall health of aquatic communities (Ouyang et al., 2002). Therefore, in view of the spatial and temporal variations in the physico-chemical, hydrological and biological attributes of surface water systems, regular monitoring programs are required for reliable estimates of the water quality.

Rivers are the only ecosystem characterized by strong and predominantly unidirectional flows of materials that intimately connect the upstream and downstream reaches (Thompson & Lake, 2010) and thus, rivers play a major role in assimilation or carrying off the municipal and industrial wastewater and nutrient removal from agricultural fields and mineral rocks by surface runoff are responsible for river pollution. The municipal and industrial wastewater discharge constitutes the constant polluting source, whereas, the surface run-off is a seasonal phenomenon, largely affected by climate in the basin (Sing et al. 2004; Vega et al., 1998). Therefore, since rivers are the most important inland water resources for human consumption, it is imperative to have reliable information on characteristics and trends of water quality for effective water management.

The usual program of water quality assessment is the periodic measurement of multiple parameters in different monitoring stations which resulted in a complex data matrix of a large number of physico-chemical parameters that should be assessed to evaluate water quality (Chapman, 1992; Dixon & Chiswell, 1996). To simplify the problem of data reduction and to draw meaningful conclusion, several researchers used water quality indices (WQI) to verify the influence of waste discharges on water quality of streams and rivers (e.g. Cao et al., 1996; Pesce & Wundelin, 2000). However, WQI are often specific to the type of pollution or the geographical area involved (Rosenbeg & Resh, 1993) and has difficulty in universal applications. In addition, they do not provide evidences on the pollution sources (Pesce & Wundelin, 2000). Similarly, univariate procedure is a common technique applied in river water quality monitoring which does not adequately characterize simultaneous similarities and differences between samples or variables (Dixon & Chiswell, 1996). In addition, the intrinsic values of analytical data are inadequate for the investigation of multivariate data table as the variables are correlated (Vega et al. 1998). Therefore, the indirect relationship between analytical parameters should be taken in to account for complete understanding of surface water quality.

Multivariate statistical methods have been widely applied in environmental data reduction and interpretation of multiconstituent chemical, physical biological measurements (Ramzakh et al., 2010; ter Braak & Verdonschot, 1995; Wenning and Erickson, 1994). These techniques have been applied for identification factors that influence water systems for reliable management of water resources as well as for rapid solution for pollution problems (Reghunath et al., 2002; Simeonov et al., 2004). In this regard, PCA is a very powerful multivariate statistical analysis method technique which is applied to reduce the dimensionality of a data set consisting of a large number of inter-related variables, while retaining as much as possible the variability present in data set (Jianqin et al. 2010; Sing et al. 2004). In addition, it allows to assess the association between variables, since they indicate participation of individual chemicals in several influence factors (Vega et al., 1998).

Since certain correlations exist among multi-indicators, PCA attempts to transform a large set of inter-correlated indicators into a smaller set of composite indicators, uncorrelated (orthogonal) variables called principal components (PCs), and simplifies the structure of the statistical analysis system (Jianqin et al. 2010). In this way, the correlation coefficient matrix measures how well the variance of each constituent can be explained by relationship with each of the others (Liu et al., 2003) and PC provides information on the most meaningful parameters, which describe the whole data set affording data reduction with minimum loss of original information (Helena et al., 2000; Vega et al., 1998). The characteristic root (eigenvalues) of the PCs is a measure of associated variances and the sum of the eigenvalues coincides with the total number of variables (Razmkhah et al. 2010). Correlation of PCs and original variables is given by loadings, and individual transformed observations are called scores (Wunderlin et al., 2001). Liu et al. (2003) classified the factor loadings as 'strong', 'moderate' and 'weak', corresponding to absolute loading values of >0.75, 0.75-0.50 and 0.50-0.30, respectively. However, loading reflects the relative importance of a variable within the component and does not reflect the importance of the component itself (Davis, 1986 cited in Ouyang, 2005).

A rotation of principal components can achieve a simpler and more meaningful representation of the underlying factors by decreasing the contribution to PCs of variables with minor significance and increasing the more significant ones (Vega et al. 1998). However, rotation might have resulted in an increase of the number of factors necessary to explain the same amount of variance of the original data set. However, it allows the association of small groups of variables and individual rotated factors with a clearer hydrochemical meaning (Vega et al., 1998) which greatly helps in data interpretation (Helena et al., 2000; Morales et al., 1999; Simeonov et al., 2003; Vega et al., 1998).

2. Application of PCA in physico-chemical water quality monitoring

The main problem in regular water quality monitoring programmes is the generation of large physico-chemical data matrix in a relatively short period of time which necessitate effective data handling mechanism for interpretation of results, association of variables and meaningful conclusion.

2.1 Anthropogenic and seasonal effects on water quality

Vega et al. (1998) applied exploratory data analysis for the assessment of the seasonal and polluting effects on water quality of Pisuerga river (Duero basin, Spain). These authors reported that the overall component loadings (i.e., no seasonal loading provided) for 22 experimental variables. PC1 explained 46.1% of the variance and PC2 explains 19.0% of the variance and it is highly participated by the variables related to anthropogenic pollution like BOD, COD, phosphorous or nitrogen. However, VF1 explained 37.2% of the total variance and is highly participated by mineral component of the river water (calcium, chloride, conductivity, dissolved solids, hardness, bicarbonate, magnesium, sodium and sulphate). VF2 contained 16.7% of the variance and include BOD, COD and ammonia. Therefore, these two verifactors identified the natural (mineral) component of pollution from anthropogenic organic pollution.

However, the study conducted by Vega et al. (1998) has not address seasonal effects which are known to have significant effect on water quality and might have affected the results of the study when interpreted for the different seasons. Ouyang et al. (2006) addressed the seasonal changes in surface water quality of the Lower Saint John River (LSJR) by the application of PCA. The authors found that PC1 explained 56.8% of the total variance measure the preponderance of physical (i.e., color, DO, and BOD) and organic-related (i.e., TKN, TOC, and DOC) water quality parameters over the mineral (i.e., alkalinity, salinity, and EC) and inorganic nutrient (i.e.,TNH_3, $DNOx$,TP, and $PO4^{-3}$) related water quality parameters while component 2 explained 26.8% of the total variance distinguished the importance of anthropogenic inputs and physical parameters (e.g., temperature and turbidity) over the natural inputs (e.g., pH, alkalinity, and salinity) during the spring season. However, unlike the cases for PC1 in spring and summer, the PC1 in fall which explained 54.2% of the total variance was positively contributed by mineral inorganic nutrient-related parameters and was negatively participated by the physical and organic-related parameters. Overall, their results revealed a high seasonal variation of water quality parameters in the dynamic river system and showed their significance (seasonal variation) when establishing the pollutant load reduction goals (PLRGs) and developing the total maximum daily loads (TMDLs). In a separate study, Ouyang (2005) applied PCA for evaluation of river water quality monitoring stations (22) in the same stream. Results showed that the first component accounted for about 94.6% and the second component accounted for about 4.5% of the total variance in the data set. PC1 is, therefore, the only one major source of data variation and 3 monitoring stations were identified as less important (non-principal) in explaining the annual variance of the data set. The authors attributed the very high variation explained by PC1 to the use of monitoring stations rather than water quality parameters in. In fact, it is expected that the water quality parameters which are controlled by hydrological, chemical and biological conditions to have higher correlations than water quality parameters which are solely controlled by hydrological conditions.

Numerous studies also confirmed that multivariate statistical techniques served as an excellent exploratory tool in understanding their temporal and spatial variations on water quality (Sing et al, 2004). The application of PCA by Razmkhah et al. (2010) discriminate the anthropogenic and "natural" influences on Jajrood river in Iran. PCA has allowed identification of a reduced number of mean 5 varifactors, pointing out 85% of both temporal and spatial changes. Rotation of the selected factors explained that VF1 (mineral contents) and VF2 (mineral and anthropogenic contamination) in spring time identified sites with worst water quality (high mineral content and higher organic pollution) and 43% of the existing variance was briefly contributed by minerals (temporal variations) whereas 26% by anthropogenic factors (spatial variations). However, PCA extracted 3 VFs for the summer season and 15.5% of the variance is contributed to by organic factors, 15.5% by minerals and 55% by both and identified the most polluted rivers. Performing PCA in the autumn season found that 39% of the variation was due to temporal factors, 10% due to organic factors and 36% was the result of both sources. PCA extracted 4 VFs for winter season and 38% was due to minerals, 12% due to organics and 31% developed by both sources. Overall, every season revealed the presence of either mineral or anthropogenic or both sources of pollution.

Recently, Fan et al. (2010) applied PCA for spatial water quality assessment and pollution sources identification in Northern, Western and Eastern part of Pearl River delta (China). The results of the PCA suggested the parameters responsible for water quality variations in North River region was mainly related to organic related parameters (DO and COD_{Mn}), inorganic nutrients (NH_3-N and TP) and metal Hg; but in East River region, it was mainly related to organic related parameters (BOD_5) and inorganic nutrients (NH_3-N and TP), and in West River Region, mainly related to organic related parameters (COD_{Mn}) and inorganic nutrients (NH_3-N and TP). Therefore, PCA offer a useful tool for assessment of water quality and management of water resources in some regions with a large number complex water quality datasets involved. Similarly, Sing et al. (2004) applied multivariate statistical techniques for evaluation of temporal and spatial variations in water quality in Gomti river (India). A varimax rotation (raw) of the PCs to six different VFs of eigenvalue > 1which are considered significant (Kim & Mueller, 1987; Liu et al., 2003) explained about 71% of the total variance. VF1 explained 17.6% of total variance and has strong positive loadings on EC, chloride, potassium and sodium and this VF represents a mineral component of the river water. VF2, explained 16.2% of total variance and has strong positive loadings on BOD and COD which represent anthropogenic pollution sources. Overall, these results from temporal PCA suggested that most of the variations is explained by the set of soluble salts (natural) and organic pollutants (anthropogenic). However, this finding is in contrast to other studies (e.g. Fan et al., 2010; Vega et al., 1998) as PCA does not result in much data reduction, as it still need 14 parameters (about 60% of the 24 parameters) to explain 71% of the data variance. Parinet et al. (2004) successfully reduced the number of analytical parameters from 18 to 4 (pH, conductivity, UV absorbance at 254 nm and permanganate index for raw water) in a study of 10 tropical lakes in Ivory Coast without notably impairing the quality of the PCA representation. However, this difference might be related to the difference in the water system (river vs. lake) and geographical factors. Overall, simplification of water quality parameters to easily quantifiable ones eases water quality monitoring programmes.

3. Changes in biological community structure

Kebede et al (2010) applied PCA to detect changes in community composition of macronvertebrates arising from wet coffee processing effluents in major coffee producing region of Ethiopia by comparing upstream sites (control sites without any impact from the effluent and other possible pollutants because of their location above processing stations) with downstream locations (locations below coffee processing stations which are effluent receivers). The relationships between the environmental and biological data were assessed using canonical multivariate analysis with the software program CANOCO 4.5 (ter Braak, Smilauer 2002). First, detrended correspondence analysis (DCA) of square-root transformed taxa abundance, with down weighting of rare taxa, detrending by segments and non-linear rescaling was used to determine the biological turnover, or gradient length, of the species data set. DCA for taxa (species) abundance of the first axis was less than 3, implying that taxa abundance exhibit linear response to environmental gradients (Leps & Smilauer, 2003). Therefore, principal component analysis (PCA) was used in the ordination of taxa abundance, and sampling sites by focusing scaling on inter-sample distances, standardization the species score (species score divided by the standard deviation), log transformation and centring by the species. In addition, to check the influence of environmental variable in explaining the variation among response variables (species), a PCA analysis of the response variable was run by using the physico-chemical data as supplementary environmental variable.

3.1 Physcico-chemical parameters

There was a highly significant variation between BOD values of the study sites (p < 0.01). BOD levels extend from 0.8 mg/l at upstream site of Urgessa river to 1900 mg/l and 1700 mg/l at downstream sites of Bore and Fite rivers, respectively. Similarly, there was a significant variation between DO values of the study sites as expected (p < 0.05). The upstream sites showed good oxygen content as the DO values were above 5mg/l. However, DO value is totally depleted at the downstream site of Bore river which is in agreement with the high BOD value recorded for the site. Conversely, a slight reduction in pH values on average (7.03 to 6.74) might be attributed to the high assimilation capacity of water. The relatively higher amount of TDS at the upper site of Chiseche river might be attributed by the high mucilage coming out from coffee processing stations which are located around Chiseche river while the variation in temperature could be related to daily temperature variation during sampling period. A general pattern of NO_3 and NH_4^+ increment at downstream sites compared the upstream sites was observed during the sampling periods. Overall, the result of the physico-chemical analysis supports similar findings such that the main ecological effect of organic pollution in a watercourse is the decrease in oxygen content (Murthy et al., 2004; von Enden & Calbert, 2002).

3.2 Descriptive analysis of macroinvertebrates

To assess the downstream water quality (river water that receive discharge from wet coffee processing stations), 6047 macroinvertebrate individuals representing 27 different taxa were collected and identified from riffle sampling sites at the upstream and downstream locations (Table 2).

Species abundance is generally believed to be a useful measure of the severity of pollution (Sheehan, 1984). The total number of individuals found at the downstream sites was 5459 which was compared to 588 individuals collected from their respective upstream sites. The highest number of individuals was found at the lower course of Guracho river (2166 individuals) while the lowest number was recorded at the upper course of Chore river with a total of 3 individuals. Almost all the sampling sites displayed higher number of individuals at the downstream sites than the upper ones (Table 3). The highest number of

Rivers	No.		BOD		DO		TDS		pH		Temp		NO$_3$-N		NH$_4^+$-N	
	U	L	U*	L*	U	L	U	L	U	L	U	L	U	L	U	L
Sugo	1	2	1.5	600	5.98	5.98	120	50	7.06	7.06	18.3	18.3	2.51	2.13	0.445	0.365
Sunde	3	4	4	330	6.82	6.54	50	245	7.17	6.56	15.4	15.3	2.28	2.51	0.322	0.406
Chiseche	5	6	1	2.3	6.6	6	295	210	6.8	7.35	17.5	18.3	0.48	1.58	0.330	0.348
Songa	7	8	270	150	5.4	5.4	195	220	6.73	6.63	17.4	17.2	1.83	2.05	0.424	0.54
Urgessa	9	10	0.8	200	7.64	7.13	145	275	7.8	6.99	16.1	16.1	1.28	2.55	0.344	1.299
Janje	11	12	1	600	5.15	6.87	95	130	6.73	7.39	19	17.4	1.64	2.09	0.208	0.431
Fite	13	14	1.3	1700	7.02	6.35	95	140	7.2	6.99	17.8	18.5	1.43	2.13	0.200	0.245
Guracho	15	16	6	5.3	6.15	5.71	60	90	6.83	6.75	18.3	18.5	0.94	1.03	0.246	0.323
Funtule	17	18	110	2.5	5.15	3.93	120	135	6.73	5.91	19	19.5	1.25	2.50	0.406	0.655
Bore	19	20	2.3	1900	6.49	0	160	255	6.96	5.87	17.2	18.1	2.41	2.80	0.365	0.604
Chore	21	22	1.9	3.2	6.04	3	120	125	7.27	6.59	18.5	19.6	1.42	1.39	0.355	0.346
Average			36.3	499.4	6.22	5.17	132.3	170.5	7.03	6.74	17.7	17.9	1.58	2.07	0.331	0.505

Table 1. Mean physico-chemical characteristics of water samples of rivers of Jimma zone at locations above (U) and below (L) coffee effluent discharge points. All units except pH and temperature (°C) are in mg/l. (Data from Kebede et al., 2010).

Family name	Code	1	2	3	4	5	6	7	8	9	10	11	12	13	14	15	16	17	18	19	20	21	22
Atericidae	ATR	-	-	-	-	-	3	1	1	2	-	-	-	-	-	1	-	-	-	-	-	1	-
Baetidae	BAE	2	-	-	4	4	2	5	6	-	1	40	15	-	1	22	6	-	-	1	1	1	1
Ceratopogenidae	CER	25	70	-	10	-	-	-	-	13	45	-	21	-	58	-	355	-	-	-	6	-	39
Chaoboridae	CHA	19	46	-	-	-	-	-	-	12	28	-	16	-	29	-	200	-	-	-	-	-	22
Chironomidae	CHR	107	871	1	42	10	3	5	11	73	157	2	224	55	506	2	1602	9	28	-	467	-	329
Corydalidae	COR	-	-	-	-	-	-	-	-	-	-	-	-	-	-	1	-	1	-	-	-	-	-
Elmide	ELM	-	-	-	-	-	-	-	-	-	-	-	-	-	-	-	-	-	-	-	-	-	-
Ephemeridae	EMP	2	-	-	-	-	-	-	-	11	-	-	-	-	-	-	-	-	-	-	-	-	-
Ephemerillidae	EPH	-	-	-	-	-	-	-	-	1	-	-	1	-	-	3	-	-	-	9	-	-	1
Fiddler Crab	FDC	-	-	-	-	-	1	1	-	-	-	-	-	-	-	-	-	-	-	-	-	-	-
Gyrinidae	GYR	1	-	-	-	-	-	-	-	-	-	-	-	1	-	-	-	1	-	1	-	-	-
Heptageniidae	HEP	-	-	-	-	-	-	1	-	1	-	1	-	-	-	-	-	-	-	-	-	-	-
Hydrophilidae	HYR	1	-	-	-	-	-	-	-	-	-	-	-	9	-	-	-	-	-	-	-	-	-
Hydropsychidae	HYD	-	-	1	-	-	13	7	7	19	23	6	3	-	-	9	-	-	1	-	1	-	-
Hydroptilidae	HYP	-	-	-	-	-	-	-	-	8	-	-	-	-	-	-	-	-	-	-	-	-	-
Libellulidae	LIB	-	-	-	-	1	1	2	-	-	-	-	-	-	-	-	-	-	-	-	-	-	-
Macromiidae	MAC	1	-	-	-	-	2	2	-	-	-	2	-	-	-	6	-	-	-	-	1	-	5
Muscidae	MUS	-	-	-	-	-	-	2	-	-	-	4	9	-	-	5	-	-	-	-	-	-	-
Perlidae	PEL	-	-	-	-	-	-	1	-	-	-	-	-	-	-	1	-	-	-	-	-	-	-
Perlodidae	PER	-	-	-	-	-	-	-	-	-	-	2	-	-	-	-	-	-	-	-	-	-	-
Planarium	PLA	-	-	-	-	-	-	-	-	-	-	-	-	-	-	-	3	-	-	-	-	-	-
Polycentropidae	POL	-	-	-	-	-	-	3	-	-	-	-	-	-	-	-	-	-	-	-	-	-	-
Psychomidae	PSY	-	-	-	-	-	-	-	2	5	-	-	-	-	-	-	-	-	-	-	-	-	-
Round worms	RW	-	-	-	-	-	8	1	5	2	-	-	2	-	-	-	-	-	1	-	-	-	-
Oligochaeta	OLI	19	131	-	5	-	-	-	-	-	-	-	1	-	-	6	-	-	-	-	-	1	-
Syrphidae	SYR	-	-	-	-	-	-	-	-	-	-	-	-	3	-	-	-	-	7	-	-	-	-
Tipulidae	TIP	-	-	2	-	-	2	1	-	-	-	-	-	-	-	3	-	-	-	4	-	-	1

Table 2. Macroinvertebrate community identified from each sampling sites with their given codes. Name of sampling points are given in Table 1. –indicates absence of the species from the site during the sampling period.

Names	Species Richness		Total Species		Maximum		Shannon Index		Equitability Index		EPT		CHR		HFBI	
	U	L	U	L	U	L										
Sugo	177	1118	9	4	107	871	0.6688	0.4056	0.4069	0.2467	5	0	107	871	7.714	8.050
Sunde	4	57	3	3	2	42	0.4515	0.323	0.2747	0.1965	1	0	1	42	4.5	7.780
Chiseche	15	35	3	9	10	13	0.3488	0.7905	0.2122	0.4809	4	15	10	3	7	4.628
Songa	32	32	13	6	7	11	1.000	0.6883	0.6085	0.4188	17	13	5	11	5.25	5.813
Urgessa	147	254	11	5	73	157	0.8225	0.5866	0.5005	0.3569	40	24	73	157	6.387	7.267
Janje	57	292	7	9	40	224	0.4757	0.5116	0.2894	0.3113	49	19	2	224	4.245	7.508
Fite	68	594	4	4	55	506	0.2775	0.3139	0.1688	0.1910	9	1	55	506	7.632	7.797
Gurracho	59	2166	11	5	22	1602	0.8485	0.4356	0.5163	0.2651	35	6	2	1602	4.695	7.655
Funtule	11	37	3	4	9	28	0.2606	0.3132	0.1586	0.1905	0	1	9	28	6.909	8.189
Bore	15	476	4	5	9	467	0.4429	0.0489	0.2695	0.0297	10	2	0	467	1.933	7.951
Chore	3	398	3	7	1	329	0.4771	0.4439	0.2903	0.2701	1	2	0	329	5.166	7.735
Total	588	5459									171	83	264	4240		
Average							0.5725	0.44192	0.3359	0.2688					5.5846	7.306

Table 3. Descriptive analysis of the macroinvertebrate community, summary of diversity, equitability and biotic indices of the study sites

species found was 13 in Songa lower with 32 individuals. The upper and lower course of Sunde river and the upper course of Chiseche and Chore rivers had only 3 species, respectively. The maximum number of individuals which belong to the same species was found at the downstream site of Gurracho river (1602 individuals of the Chironomidae family). The general increment in total number of individuals at the downstream sites might be attributed to the enrichment of downstream sites from coffee processing wastewater, which provides physiologically adapted organisms to exploit the excess nutrients available despite low oxygen tension and this response of macroinvertabrates to organic pollution has been well documented (Cao et al., 1996; Hilsenhoff, 1988; Johnson et al., 1993). However, the general increment in species richness at the downstream sites was not supported with species diversity. Hence, the downstream site of Gurracho river which had 2166 individuals were found to contain only 5 species while the upstream sites of Chore river which had 3 individuals were belong to different species. This finding is consistent with Johnson et al. (1993) in that pollution tolerant species like Chironomidae species dominate the impacted sites.

3.3 Biotic and diversity indices

In order to understand the effect of wet coffee processing discharge on the biotic environment of the rivers, different diversity indices were tested (Table 3). These indices would indicate the environmental impact of coffee processing activities on the surrounding environment. The highest Shannon index was found at the upstream site of Songa river while the lowest Shannon index was found at the downstream sites of Bore and Urgessa river (Table 3). Similarly, Shannon index decreased from 0.57 at the upstream sites to 0.44 at the downstream sites. But the Equitability index which describes the evenness of species distribution within the site showed a different pattern. The highest and lowest Equitability indices were found at the upstream and downstream sites of Songa and Bore rivers, respectively. High values of diversity indices and equitability at the downstream sites of Songa river despite its presence below coffee discharge points, were unimpacted by coffee processing effluent. This finding of biological data is also supported by physicochemical data. Its good river water quality might be related to efficient coffee effluent control mechanism (lagoon) and the relative placement of the stations at a higher distance from the rivers i.e. large area possessed by the station enabled the construction of efficient pits to the containment of both coffee wasterwater and pulp. In general, Equitability index decreased from 0.336 at the upstream sites to 0.268 at the downstream sites and more than 50% of the sites had an Equitability index of less than 0.3 and, thus, few species had dominated these sites.

The highest number of Ephemenoptera, Plecoptera and Tricoptera (EPT) individuals was found at the upper course of Janje river followed by Urgessa river, while the lowest EPT taxa were found in almost all of the downstream sites and upstream site of Funtule rivers. The total number of EPT taxa at the upstream sites was 171 and declined to 83 at the downstream sites. However, the number of Chironomidae (CHR) sharply increased from 264 at the upstream sites to 4240 at the downstream sites (Table 3), which is about 16 times higher than the upstream sites. The low diversity at the upstream sites of Fite and Funtule rivers was mainly related to the dominance of these sites by Chironomidae family. It comprises more than 76% of all the individuals found in at these sites. The situation is also the same at the upper course of Funtule river. It had three species which was highly dominated by Chironomidae taxa. Many reports have shown a significant decrease in species diversity indices associated with pollution

(Norris & Georges, 1993; Sheehan, 1984). Similarly the high EPT taxa at the upstream sites of Urgessa and Janje is related to their low BOD values.

With regard to the Hiselhof Family Biotic Index (HFBI) evaluation, the upper course of Bore (1.933) river had the lowest HFBI score; whereas Funtule river had the highest HFBI value followed by the lower course of Sugo rivers. Similarly, the family biotic index increased from 5.585 at the upstream sites to 7.306. In fact 14 of the 22 downstream sites had HFBI values above 6.51. . The abundance of pollution tolerant Oligochaeta taxa and Syrphidae species at the downstream sites of Sugo and Funtule rivers resulted in high FBI score. Blood-red chironomids and other dipterans (e.g., Ceratopogenidae and Chaoboridae) that are considered indicators of severely polluted sites were found at the downstream sites in high abundance than upstream sites. In fact, a benthic community dominated by one or few taxa is often indicative of environmental stress.

3.4 Principal component analysis

The separate analysis of environmental variables and biological indicators resulted in a slight difference results in depicting the pollution gradient along the study rivers. In fact this is expected since the abundance of species at a given site is more a reflection of past environmental conditions (Richard et al., 1997), while measurements of physical and chemical factors may be more of an indication of present conditions. However, the combination of both environmental variables (Table 1) and biological indicators (Table 2) by PCA was able to clearly define and explain the pollution gradients.

3.4.1 Community composition of macroinvertebrates

Based on the species data of the PCA analysis (Fig.1), the first axis described 56.7% of the variation in the species composition while the second axis described 11.7% of the variation. Therefore, the first two axes were responsible for 68.4% of the variation in the dataset.

The dissimilarity between sites (12-2) dominated by Chaoboridae (CHA), Ceratopogenidae (CER), Chironomidae (CHR), Oligocheata (OLI) on the right side of the first axis with those sites (15-8) at the top left of the diagram was high as shown by the distance in the PCA analysis.

The distribution of Chaoboridae, Ceratopogenidae Chironomidae were centered at sites 10 (Urgessa lower), 22 (Chore lower), 14 (Fite lower), 2 (Sugo lower) and 16 (Gurracho lower). The distribution of Chironomidae, the most abundant taxa, at the downstream sites indicated the impact of coffee processing effluent on the community compositon of these particular sites. Similarly, many reports have documented a significant increase by Chironomidae taxa following toxic exposure (eg. Clements, 1994). On the contrary, Hydropsychidae (HYD), Baetidae (BAE), Heptagenidae (HEP), Perlidae (PER), and Psychodidae (PSY) was found at the negative of the first axis. These species have been used as indicators of good water quality (Cao et al., 1996; Legesse, 2001). In general, the PCA rendered three classifications: the extremely impacted downstream sites at the right side of the first axis and the two classes located at the right and negative side of the second axis which mostly constitute the upstream sites and their separation is attributed to their difference in species composition.

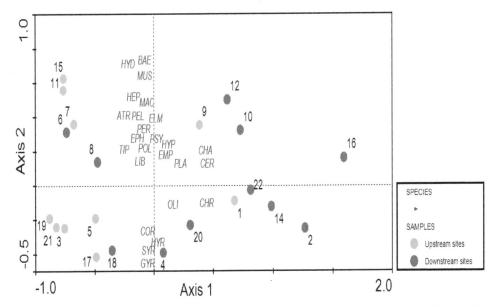

Fig. 1. PCA biplot of samples and macroinvertebrates based on the first two axes. Name of sampling points are given in Table 1 and full name of abbreviation codes of macroinvertebrate taxa is given in Table 2.

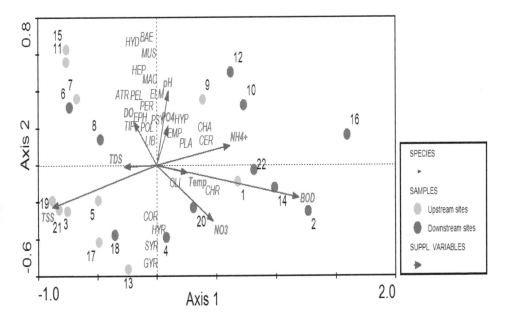

Fig. 2. PCA triplot of samples, macroinvertebrates and environmental variables based on the first two axes.

The species and environmental variables together described about 41% of the variation in the data set and the first two axes are responsible for 70% of these variations or 28.7% of the total variation (70% X 41%) suggesting that the relative importance of the first two axes in describing the variation in taxa separation and environmental variables.

In the ranking of the projection points, the origin (0, 0) indicates the global average of the variable. Therefore, TDS, Temperature and PO_4 did not change much (short arrow length) and hence their influence were minimal as compared to BOD, TSS, NO_3, pH, and DO environmental lines which displayed the maximum rate of change across the diagram. Therefore, variations in macroinvertebrate composition were strongly correlated with these environmental variables. The high oxygen weighted average for oxygen loving species of Hydropsychidae (HYD), Baetidae (BAE), Heptagenidae (HEP), Perlidae (PER) and Psychodidae (PSY) is as expected since Heptagenidae is in Ephemeroptera family while Hydropsychidae, Perlidae and Psychodidae are in Trichoptera family. Ephemeroptera, Trichoptera and Plecoptera are considered to be abundant in oxygen rich water, very sensitive for pollution and used as a water quality monitoring index in Ethiopia (Kebede et al., 2010; Legesse, 2001). Therefore sites 8 (Sugo lower), 6 (Chiseche lower), 7 (Songa upper), 11 (Janje upper), and 15 (Gurracho upper) had higher DO values than the rest sites.

On the contrary, organically polluted sites were found to be dominated by pollution tolerant species of Chaoboridae (CHA), Oligochaeta (OLI), Ceratopogenida (CER) and Chironomidae (CHR) taxa and except the taxa of Oligochaeta, all the species are in the Diptera family. The special characteristics of the above mentioned taxa are their ability to flourish in very low oxygen (high BOD) level and dominate the entire site with great abundance. Sites 14 (Fite lower) and 2 (Sugo lower) and 20 (Bore lower) had the highest BOD values than the rest of the sites. Therefore the species-site-environment PCA triplot diagram clearly discriminated unimpacted sites from the severely impacted ones.

3.4.2 Sporulation pattern of aquatic hyphomycetes

Recently, Kebede applied PCA for discrimination of the sporulation pattern of aquatic hyphomycetes (aquatic fungi) communities on oak leaves immersed in a low order stream for up to 12 weeks in Candal stream (Central Portugal; 40° 4' 44" N and 8° 12' 10" W). In this study, the significance of the sporulation temperature at the peak of sporulation (Table 4), day 28 at 20°C (T4_20), day 42 at 10°C (T6_10) & 15°C (T6_15) and day 56 at 5°C (T8_5), was evaluated using PCA.

The non metric dimensional scaling (MDS) was applied in Winkyst program by using the Bray-Curtis distance matrix for producing the species axes (coordinates) that were used as a species data in principal component analysis (CANOCO 4.5; ter Braak & Smilauer, 2002).

The response of the dominant species at the four temperature values is shown in Fig. 3. The stress value (fit of regression) of 0.038 depicted the excellent representation of the coordinate axes in discriminating the samples (sampling dates and incubation temperature) in the ordination diagram. The supplementary species data indicate the species of *Clavariopsis aquatica* (AQU) and *Mycocentrospora acerina* (ACE) were associated with the discrimination of the positive side of the axis (right side) by over-dominating the sporulation at 5°C.

Conversely, the species of *Tetrachaetum elegans* (ELE) and *Articulospora tetracladia* (TET) and to a lesser extent *Alatospora acuminata* (ACU) and *Clavatosppora longibrachiatum* (LON) were associated with the discrimination of the negative side of the axis (left side). These species highly dominate the sporulation at 10, 15 and 20°C. In addition, the PCA diagram shows a pattern of species dominance (longer arrow length) mainly due to ELE and ACE during the study period. Overall, the PCA analysis showed the species specific response to temperature variation by the dominant species. This is particularly pronounced during the peak period of sporulation with ACE, AQU, ELE and TET preferentially sporulate at 5°C, 10°C, 15°C and 20°C, respectively. This pattern revealed by PCA appears to be due primarily to the differential response by the late colonizer ACE and early colonizer ELE to low and high temperatures, respectively. This finding adds to similar changes in community composition induced by temperature (Bärlocher & Kendrick 1974, Suberkropp, 1984; Gessner et al. 1993; Bärlocher et al. 2008; Dang et al., 2009; Fernandes et al., 2009; Ferreira & Chauvet, 2011).

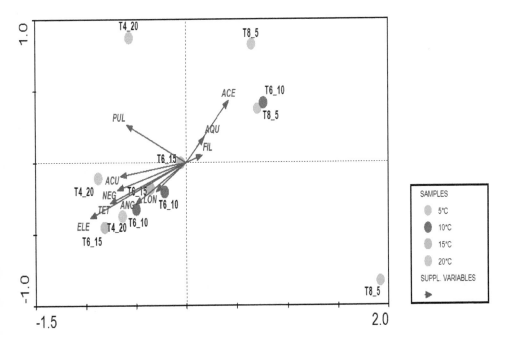

Fig. 3. PCA diagram of the sporulation temperature constructed on the species axes. The original species data was used as a supplementary variable.

Species	T8_5			T6_10			T6_15			T4_20		
	1	2	3	1	2	3	1	2	3	1	2	3
Anguillospora filiformis (FIL)	0	0	3	1	2	0	0	0	0	0	0	0
Alatospora acuminate (ACU)	5	0	1	22	32	1	0	22	7	12	65	10
Alatospora pulchella (PUL)	0	0	0	0	0	0	3	0	2	8	5	0
Articulospora tetracladia (TET)	0	1	1	3	13	0	15	8	8	4	32	40
Clavariopsis aquatica (AQU)	33	3	24	17	17	33	3	15	79	0	9	13
Clavatosppora longibrachiatum (LON)	4	0	5	60	48	4	13	7	5	0	0	0
Mycocentrospora acerina (ACE)	70	1	27	0	2	0	0	2	0	0	0	0
Stenocladiella neglecta (NEG)	0	0	0	0	0	0	5	1	0	2	1	5
Tetrachaetum elegans (ELE)	10	0	13	89	38	10	157	83	58	17	106	134
Tricladium angulatum (ANG)	0	0	0	0	0	0	1	0	0	0	0	1

Table 4. Abundance of aquatic hyphomycetes conidia associated with oak leaves during the peak period of sporulation at three replications. T4, T6 and T8 represent 4, 6 and 8 weeks of incubation period in weeks while 5, 10, 15 and 20 represent incubation temperatures (°C).

4. Conclusions

The usual program of water quality assessment depends on the periodic measurement of multiple parameters in different monitoring stations which resulted in a complex data matrix of a large number of physico-chemical parameters. Therefore, to simplify the problem of data reduction and draw meaningful conclusion multivariate statistical techniques have been widely used. In this regard, PCA, a powerful multivariate statistical technique, is applied to reduce the dimensionality of a data set while retaining as much as possible the variability present in data set and allows to assess associations between variables. PCA has been applied for the assessment of the seasonal and polluting effects on water quality, evaluation of monitoring stations, temporal and spatial variations on water quality and pollution source identification with in the river basin. Recently, PCA was applied to assess the water quality of rivers impacted by organic pollution. The combination of both environmental variables and biological indicators in PCA was able to clearly define and explain the pollution gradients. Similarly, PCA enabled to identify the distinct sporulation pattern of aquatic hyphomycetes communities incubated at four different temperatures. Therefore, the use of PCA to detect changes in community structure, ecological integrity of surface water systems and environmental impact assessment will enable to understand an integrative health condition of surface water systems due to chemical, physical and biological stressors.

5. Acknowledgments

I would like to thank Professor Felix Bärlocher and Professor Cristina Canhoto for their guidance and constructive comments during the sporulation study in Portugal. I am also grateful to Dr. Fassil Assefa for his valuable comments and guidance during the study in Jimma zone, Ethiopia.

6. References

Bärlocher, F. & Kendrick, B. (1974). Dynamics of fungal populations on leaves in a stream. *Journal of Ecology*, Vol.62, No. 3, pp. 761-791.

Bärlocher, F.; Seena, S.; Wilson, K. & Williams, D. (2008). Raised water temperature lowers diversity of hyporheic aquatic hyphomycetes. *Freshwater Biology*, Vol. 53, No. 2, pp. 368-379.

Cao, Y.; Bark, A. & Williams, W. (1996). Measuring the responses of macroinvertebrate communities to water pollution: a comparison of multivariate approaches, biotic and diversity indices. *Hydrobiologia*, Vol. 341, No. 1, pp. 1-19.

Chapman, D. (1992). *Water Quality Assessment; A guide to the use of biota, sediments and water in environmental monitoring*. University Press, Cambridge.

Clements, W. (1994). Benthic invertebrate community responses to heavy metals in the Upper Arkansas River Basin, Colorado. *Journal of the North American Bentholological Society* Vol. 13, No. 1, pp. 30-44.

Dang, C.; Schindler, M.; Chauvet E. & Gessner, M. (2009). Temperature oscillation coupled with fungal community shifts can modulate warming effects on litter decomposition. *Ecology* Vol 90, No. 1, pp. 122-131.

Dixon, W. & Chiswell, B. (1996). Review of aquatic monitoring program design. *Water Research*, Vol. 30, No. 9, pp. 1935–1948.

Fan, x.; Cui, B.; Zhao, H.; Zhang, Z. & Zhang, H. (2010). Assessment of river water quality in Pearl River Delta using multivariate statistical techniques, *International Society for Environmental Information Sciences 2010 Annual Conference (ISEIS), Procedia Environmental Sciences*, Vol. 2, pp. 1220–1234.

Fernandes, I.; Uzun, B.; Pascoal, C. & Cássio, F. (2009). Responses of aquatic fungal communities on leaf litter to temperature change events. *International Review of Hydrobiology*, Vol. 94, No. 4, pp. 410-418.

Ferreira, V. & Chauvet, E. (2011). Future increase in temperature more than decrease in litter quality can affect microbial litter decomposition in streams. *Oecologia*, Vol. 64, No. 1, pp. 279-291.

Gessner, M.; Thomas, M.; Jean-Louis, A. & Chauvet, E. (1993). Stable successional patterns of aquatic hyphomycetes on leaves decaying in a summer cool stream. *Mycological Research*, Vol. 97, No. 2, pp. 163-172.

Helena, B.; Pardo, R.; Vega, M.; Barrado, E.; Fernandez, J. & Fernandez, L. (2000). Temporal evolution of groundwater composition in an alluvial aquifer (Pisuerga River, Spain) by principal component analysis. *Water Research*, Vol. 34, No. 3, Pp. 807–816.

Hilsenhoff, W. (1988). Rapid field assessment of organic pollution with a family-level biotic index. *Journal of the North American Benthological Society*, Vol. 7, No. 1, pp. 65-68.

Jianqin, M.; Jingjing, G. & Xiaojie, L. (2010). Water quality evaluation model based on principal component analysis and information entropy: Application in Jinshui River. *Journal of Resouces and Ecology*, Vol. 1, No. 3, pp. 249-252.

Johnson, R.; Wiederholm, T. & Rosenberg, D. (1993). Freshwater biomonitoring using individual organisms, populations, and species assemblages of benthic macro- - invertebrates, In: *Freshwater biomonitoring and benthic macroinvertebrates*, Rosenberg, D.& Resh, V., pp (40-158), Chapman & Hall, New York.

Kebede, Y.; Kebede, T; Assefa, F. & Amsalu, A. (2010). Environemntal impact of coffee processing effluent on the ecological integrity of rivers found in Gomma woreda of Jimma zone, Ethiopia. *Ecohydrology and Hydrobiology*, Vol. 10, No. 2, pp. 259-270.

Kim, J. & Mueller, C. (1987). *Introduction to factor analysis: what it is and how to do it. Quantitative Applications in the Social Sciences Series*. Sage University Press, Newbury Park.

Legesse, W. (2001). *Senstivity of biotic indices to natural disturbances*. PhD thesis, Departement of Zoology and Animal Ecology, University College Cork, Republic of Ireland.

Leps, J. & Smilauer, P. (2003). *Multivariate analysis of ecological data using CANOCO*, Cambridge University Press, New York.

Liu, C.; Lin, K. & Kuo, Y. (2003). Application of factor analysis in the assessment of groundwater quality in a blackfoot disease area in Taiwan. *Science of the Total Environment*, Vol. 313, No. 1-3, pp. 77–89.

Morales, M.; Marti, P.; Llopis, A.; Campos, L. & Sagrado, S. (1999). An environmental study by factor analysis of surface seawaters in the gulf of Valencia (Western Mediterranean). *Analytica Chimica Acta*, Vol. 394, No. 1, pp. 109-117.

Murthy, K.; D'Sa, A. & Kapur, G. (July 2011). An effluent treatment-cum-electricity generation option at coffee estates: is it financially feasible?, In: *International Energy Initiative, Bangalore*.< http://iei-asia.org/IEIBLR-Clean-Coffee.pdf >

Norris, R. & Georges, A. (1993). Analysis and interpretation of benthic macroinvertebrate survey. In: *Freshwater biomonitoring and benthic macroinvertebrates* Rosenberg, D.& Resh, V., pp. (234-286), Chapman & Hall, New York.

Ougang, y. (2005). Evaluation of river water quality monitoring stations by principal component analysis. *Water Research*, Vol. 39, No. 12; pp. 2621-2635.

Ouyang, T.; Zhu Z. & Kuang, Y. (2006). Assessing impact of urbanization on river water quality in the Pearl River Delta Economic Zone, China. *Environmental Monitoring Assessment*, Vol, 120, No. 1-3, pp. 313-325.

Ouyang, Y.; Higman, J.; Thompson, J.; O'Toole, T. & Campbell, D. (2002). Characterization and spatial distribution of heavy metals in sediment from cedar and ortega rivers Basin. *Journal of Contaminant. Hydrology*, Vol. 54, No. 1-2, pp. 19–35.

Parinet, B.; Lhote A. & Legube, B. (2004). Principal component analysis: an appropriate tool for water quality evaluation and management–application to a tropical lake system. *Ecological Modelling*, Vol. 178, No. 3-4, pp. 295–311.

Pesce, S. & Wunderlin, D. (2000). Use of water quality indices to verify the impact of Córdoba City (Argentina) on Suquía River. *Water Research* Vol. 34, No. 11, pp. 2915-2926.

Razmkhah, H.; Abrishamchi, A. & Torkian, A. (2010). Evaluation of spatial and temporal variation in water quality by pattern recognition techniques: A case study on Jajrood River (Tehran, Iran). *Journal of Environmental Management*, Vol. 91, No. 4, pp. 852-860.

Reghunath, R.; Murthy, T. & Raghavan, B. (2002). The utility of multivariate statistical techniques in hydrogeochemical studies: an example from Karnataka, India. *Water Research*, Vol. 36, No. 10, pp. 2437-2442.

Richard, S.; Thorne, J. & Williams, W. (1997). The response of benthic macroinvertebrates to pollution in developing countries: a multimetric system of bioassessment. *Fresh water biology*, Vol. 37, No. 3, pp. 671-686.

Rosenberg, D. & Resh, V. (Eds). (1993). *Freshwater Biomonitoring and Benthic Macroinvertebrates* Chapman and Hall, New York.

Sheehan, P. (1984). Effects on community and ecosystem structure and dynamics, In: *Effects of pollutants at the ecosystem level*, Sheehan, P.; Miller, D.; Butler, G. & Bourdeau, pp. (51-100), John Wiley & Sons, Chichester, New York.

Simeonov, V.; Stratis, J.; Samara, C.; Zachariadis, G.; Voutsa, D.; Anthemidis, A.; Sofoniou, M. & Kouimtzis, T. (2003). Assessment of the surface water quality in Northern Greece. *Water Research*, Vol. 37, No. 17, pp. 4119-4124.

Singh, K.; Malik, A.; Mohan, D. & Sinha, S. (2004). Multivariate statistical techniques for the evaluation of spatial and temporal variations in water quality of Gomti River (India)-a case study. *Water Research*, Vol. 38, No. 18, pp. 3980-3992.

Suberkropp, K. (1984). Effect of temperature on seasonal occurrence of aquatic hyphomycetes. *Transactions of the British Mycological Society*, Vol. 82, No. 1, pp. 53-62.

ter Braak, C. & Smilauer, P. (2002). *CANOCO reference manual and CanocoDraw for windows user's guide: Software for canonical community ordination (Version 4.5)*. Microcomputer power Ithaca, New York.

ter Braak, C. & Verdonschot, P. (1995). Canonical correspondence analysis and related multivariate methods in aquatic ecology. *Aquatic Sciences*, Vol. 57, No. 3, pp. 255-289.

Thompson R. & P. Lake (2010). Reconciling theory and practise: the role of stream ecology. *River Research and Applications*, Vol. 26, No. 1, pp. 5-14.

Vega M., Pardo R., Barrado E. & L. Deban (1998). Assessment of seasonal and polluting effects on the quality of river water by exploratory data analysis. *Water Research*, Vol. 32: No. 12, pp. 3581-3592.

Von Enden, J. & Calvert, K. (June 2010). Review of coffee waste water characteristics and approaches to treatment. PPP Project, In: *Improvement of Coffee Quality and Sustainability of Coffee Production in Vietnam*, German Technical Cooperation Agency (GTZ), Available from
http://coffee.20m.com/CoffeeProcessing/CoffeeWasteWater.pdf >

Wenning, R. & Erickson, G. (1994). Interpretation and analysis of complex environmental data using chemometric methods. *Trends in Analytical Chemistry*, Vol. 13, No. 10, pp. 446-457.

Wunderlin, D.; Diaz, M.; Ame, M.; Pesce, S.; Hued, A. & Bistoni, M. (2001). Pattern recognition techniques for the evaluation of spatial and temporal variations in water quality, a case study: Suquia river basin (Cordoba Argentina). *Water Research*, Vol. 35, No. 12, pp. 2881-2894.

5

Applications of PCA to the Monitoring of Hydrocarbon Content in Marine Sediments by Means of Gas Chromatographic Measurements

Mauro Mecozzi*, Marco Pietroletti, Federico Oteri and Rossella Di Mento
Laboratory of Chemometrics and Environmental Applications, ISPRA, Rome, Italy

1. Introduction

The application of Principal Component Analysis (PCA) in biochemical studies lies in the field of Chemometrics, the discipline which describes and applies statistical multivariate methods to the laboratory studies. PCA like Cluster Analysis (CA), belongs to the so called unsupervised pattern recognition methods, multivariate methods which can be applied to any data set without requiring or supposing any preliminary knowledge about the information present in the data (Massart & Kauffman, 1983; Brereton, 2003).

PCA has been also defined "a data reduction form" for its peculiar ability to reduce the dimension of an experimental data set without loosing the qualitative and quantitative information present (Brereton, 2003). In matrix notation, the PCA decomposition of a multivariate experimental data set including several samples and called X, is reported in the equation 1

$$X = SV' + E \tag{1}$$

where the S term is the score matrix, V' is the transposed loading matrix and E is the noise matrix . With respect to the original X (n-sample, t-variables) set, the dimension of the new matrices is changed; S has (n-sample, p) dimension, V has (p, t-variables) dimension and E only retains the same dimension of X obviously. The term " p " of S and V matrices represents the number of significant principal components or factors determined by PCA; they have the peculiar ability of describing a high fraction of the total variance (i.e. information) present in the X matrix and very important, the "p" dimension is always significantly lower than the " t " dimension of the original variables of the X matrix.

This data reduction ability of PCA is very helpful when large size of multivariate data sets have to be analyzed and interpreted. In common environmental monitoring studies PCA is applied in the analysis of discrete multivariate data when for instance, several sites with their pollutant loads have to be analysed and compared (Cicero et al., 2001; Conti & Mecozzi, 2008). However in environmental studies, the power of PCA becomes even more helpful when large size set of analytical signals such as GC chromatograms have to be

---

* Corresponding Author

analyzed. In fact, gas chromatography is a widespread technique for the monitoring of oil spills in terrestrial and marine environments (Wang et al., 1999) and in the case of marine sediments, gas chromatography tries to establish several aspects concerning total hydrocarbon content and distribution for testing homogeneity and or heterogeneity of pollutant loads and for identifying the sources of oil spills (Wang et al., 1999). In any case, this last task can be hardly obtained because any chromatogram is a multivariate sample where many hydrocarbons are usually present. A typical GC chromatogram, reported in Figure 1, is a data file with 2 columns, the acquisition time of the analytical signals and their detected intensities respectively. Here, the present hydrocarbons are identified by means of their retention time (i.e. the time corresponding to the maximum peak intensity).

The chromatogram of Figure 1 shows the presence of more than fifty hydrocarbons and in addition, the fast sampling signal causes the presence of a not negligible noise which corrupts the real intensity of signals (Mecozzi & Tomassetti, 2007; Kokaly et al., 2001). As a consequence, we can hardly perform a numerical and visual comparison of different chromatograms when we try to establish homogeneous or heterogeneous hydrocarbon compositions among samples as shown by the example of Figure 2.

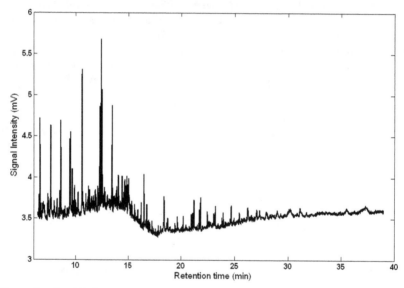

Fig. 1. Example of a GC chromatogram arising from the analysis of hydrocarbons extracted by a marine sediment. Any detected peak represents a hydrocarbon present in the sample.

According to Equation 1, PCA re-describes the starting X data set by means of a set of the new "p" variables (i.e. factors), which being significantly lower than the number of the original variables, allow to compare samples by means of simple two or three dimensional plots, using the score matrix. In addition, PCA examines the variables which determine similarity or dissimilarity among samples by means of the loading analysis. Loadings are the statistical weights of the original "t" variables of the X matrix and in the case of chromatographic data their analysis allows to identify the hydrocarbons which characterise any samples. This is a peculiar advantages of PCA with respect Cluster Analysis, that is

Applications of PCA to the Monitoring of Hydrocarbon Content in Marine Sediments by Means of Gas
Chromatographic Measurements

85

known as a fast screening method to determine similarity in experimental data set but in any case, it allows neither to determine the statistical weight of the variable nor to study peculiar variables determining qualitative similarities and dissimilarities among samples (Brereton, 2003).

However, the application of PCA to large size data set requires some necessary preprocessing treatments so to avoid potential misinterpretation of its results. In fact, a GC data file. such as the chromatogram of Figure 1, consists of about 20.000 analytical signals and when we examine a data file including thirty or forty samples, the resulting X matrix has high data dimension and redundancy. This causes high time for PCA computation and analytical problems such as reduction of the signal to noise (S/N) ratio and baseline drift. The selection of proper preprocessing treatments of chromatograms can solve all these problems and supports the correct application of PCA to large size multivariate data.

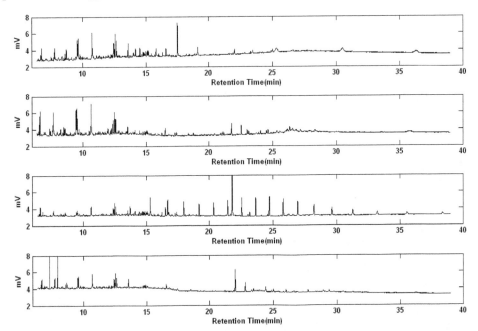

Fig. 2. GC chromatograms of hydrocarbons extracted by some sediments sampled along the Italian coasts. The simple visual examination of peak positions in the four plotted chromatograms shows how the related samples can be hardly compared for establishing similarities and dissimilarities of composition.

In this paper we discuss the application of PCA for performing the hydrocarbon monitoring in two different GC chromatographic sets. Our study takes into account all the steps for a correct application of PCA to high dimension chromatographic data files. The first set consists of 29 superficial sediments from two different areas along the coasts of Italian seas, seventeen from Venice lagoon (Adriatic sea) and twelve from Bagnoli (near Naples, Tyrrhenian sea) respectively; the second set consists of 39 subsamples of marine sediments coming from a sediment core taken in Antarctic sea.

The main purpose of PCA application is that to retrieve information hardly detectable by means of conventional methods of GC analysis of hydrocarbons in environmental studies.

2. Experimental section

This experimental study consists of five different steps; sampling of marine sediments, hydrocarbon extraction and purification from other lipid compounds present in marine sediments (Mecozzi et al., 2011), gas chromatographic analysis of the extracts, chemometric pretreatment of chromatograms and application of PCA. PCA was applied to the two different chromatographic data matrices including all the samples from the Italian coasts and the Antarctic sediment core.

2.1 Sampling of marine sediments

Marine sediment sampling from the Italian coasts was performed by a box corer, taking the upper 5 cm layer. Figure 3 reports the location of the two sampling areas along the Italian coasts. Samples were stored frozen at -25°C until chemical analysis.

Fig. 3. Map of Italian coasts showing the two areas where surface sediments were sampled. The white arrows shows the area of the Venice Lagoon in Northern Adriatic sea and the grey one shows the area of Bagnoli near Naples in Tyrrhenian Sea.

The Antarctic sediment core was sampled in the B5/Y5 station (75° 04′ South, 164° 13′ East) in the Ross bay at 550 meter of depth. This area is characterised by an intense stratification of sediment and of biogenic organic materials. The sediment core was taken by means of dredge sampler and the core was stored frozen at -25°C until GC analysis.

2.2 Hydrocarbon extraction

Hydrocarbon content was extracted and purified by means of an ultrasound method developed in our laboratory (Mecozzi et al., 2011). Each sediment sample (20 g) was added with n-hexane (20 ml) and H_2O (40 ml) at pH 2 obtained by adding concentrated HCl. Sediment was sonicated in an ultrasound cleaning bath operating at 35 kHz for 20 minutes at room temperature. Then the supernatant was separated from sediment by centrifugation. The separation of the aqueous phase from the organic phase was performed in a separating funnel; then he organic phase was dried on anhydrous Na_2SO_4. This process was repeated other twice, the extracts joint together and the organic phase was concentrated under vacuum down to 1 ml of final volume for GC analysis.

2.3 Gas chromatographic analysis

The determinations of hydrocarbons extracted by marine sediments were performed using a Carlo Erba (Milano Italy) instrument with flame ionization detector. The apparatus was equipped with a capillary GC Column Therm 1 (Thermo Scientific Milano Italy), 30 m length, i. d. 0.22 mm. Experimental conditions were injector 320°C, FID detector 360°C and the introduction was performed in spleatless mode (one minute). The temperature program used for chromatographic separation of hydrocarbons was 70°C for four minutes, thermal gradient 15°C min^{-1} to 340°C; This temperature was finally held for fourteen minutes. Chromatograms were saved as ASCII files for any further elaboration.

2.4 Chemometric pretreatments of chromatograms prior to PCA application

2.4.1 Improvements of analytical quality data and reduction of computation time

Handling of large data set prior to PCA application requires the preliminary solution of several drawbacks; in fact, the high frequency sampling of analytical signals produces data redundancy, high time of computation, with in addition analytical drawbacks such as reduction of the signal to noise (S/N) ratio and baseline drift (Christensen and Tomasi (2007). The same authors suggested several chemometric procedures for reducing these effects prior to apply PCA to GC data; with this aim, an in house MATLAB (Natik, USA) routine was applied to any collected chromatogram. In the appendix we report a MATLAB routine according to the algorithms described by Christensen and Tomasi (2007). Figure 4 reports an example of this approach. After this pretreatment, GC chromatograms were saved again as ASCII files.

2.4.2 Standardisation of the GC data set

Standardisation, also called scaling, is another fundamental step prior to PCA application, necessary for reducing the effect of the different magnitude of intensity variations in the case of multivariate data, causing uncorrected determination of the total variance of the data system (Brereton, 2003; Wang et al., 1999; Noda, 2008).

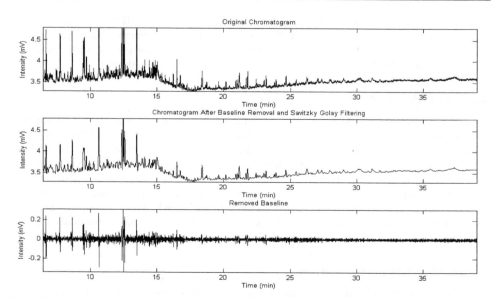

Fig. 4. Example of chemometric pretreatment of a GC chromatogram., Original chromatogram, upper plot; chromatogram with data redundancy reduction, smoothed signals and background correction, middle plot; bottom plot, removed baseline.

In environmental monitoring, where PCA is often applied to study the distribution of pollutant loads, a common scaling technique is autoscaling; given the Y column vector to be included in the X matrix and having n-sampled analytical signals, autoscaling performs the data transformation according to

$$Y_{ias} = (Y_i - Y_M)/\sigma \qquad (2)$$

where Y_i, Y_{ias}, Y_M and σ are the original value i_{th} value, its autoscaled term, the average value of the Y vector and standard deviation of the Y vector respectively. After autoscaling, any new Y series to be included in the X matrix has mean value 0 and variance value 1.

This is a very powerful approach to reduce the effect of different size ranges on the total variance of discrete data set but when applied to other types of variables such as the cases of analytical signals, autoscaling has a marked drawback. In fact, digitised files of spectroscopic and chromatographic data generally consist of several thousands of signals sampled with high frequency acquisition. In this case autoscaling can often produce the enhancement of noise depending on its division by a small value of standard deviation (Noda, 2008; Kokalj et al., 2011).

Other scaling techniques are available for solving the disadvantage originating from autoscaling. In the mean centred technique data are scaled according to

$$Y_{imc} = (Y_i - Y_M) \qquad (3)$$

where Y_{imc} is mean centred scaled value of the Y series while Y_i and Y_M are the same meaning of the equation 2.

Normalization scaling consists of transforming data according to

$$Y_{inorm} = (Y_i - Y_{min})/(Y_{max} - Y_{min}) \qquad (4)$$

where Y_{inorm}, Y_{min} and Y_{max} are the normalized Y_i term, the minimum and the maximum values of the Y series respectively. After normalization, all the Y vectors range between 0 and 1.

Pareto scaling is a technique proposed by the Italian economist Vilfredo Pareto (Noda, 2008); it consists of the division of the Y series values by the square root of its standard deviation according to

$$Yi_p = Y_i/\sqrt{\sigma} \qquad (5)$$

where Y_{ip} is the Pareto scaled of the original Yi value and σ has the same meaning of equation 2.

Any scaling technique produces different effects on the quality of analytical signals so that the selection of the opportune scaling needs a carefully evaluation of the produced results. We report examples of application of all the above scaling methods in Figures 5 and 6 so to support the selection of the most appropriate methods prior to PCA application to GC data. With respect to the original chromatogram, autoscaling causes baseline drift with negative analytical signals and in addition, noise is enhanced in some zones of the chromatogram as shown by the example of Figure 5 (middle plot).

Mean centred scaling causes a baseline drift with negative analytical signals as well, though it does not cause a S/N ratio reduction as observed for autoscaling instead (Figure 5, bottom plot).

Normalization and Pareto scaling techniques do not cause negative baseline drifts and evident noise enhancements (Figure 6) so that we recommend to apply one of these as scaling pretreatments. These techniques can be applied by means of a common spreadsheet such as Excel for Windows. In any case, in the Appendix section we report two ad hoc routines written in MATLAB language for applying the above scaling techniques.

2.5 Application of PCA to gas chromatographic data set

PCA was applied to GC chromatograms by an in house routine written in MATLAB (Natik, Wi, USA, ver 5.0) language according to the singular value decomposition algorithm described by Geladi (2002). The list of the routine is reported in the Appendix section.

2.6 Chemical reagents

All the chemical reagents used for the experimental work were of analytical reagent grade (Carlo Erba, Milan, Italy) and only ultrapure MilliQ water was used for any chemical treatments of samples.

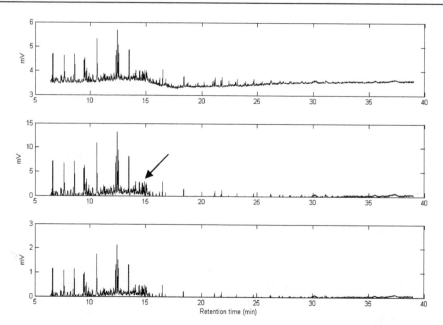

Fig. 5. Scaling methods applied to GC set. Conventional chromatogram, upper plot ; autoscaling, middle plot; mean centred scaling, bottom plot. The arrow shows a case where autoscaling increases noise with respect to the original plot.

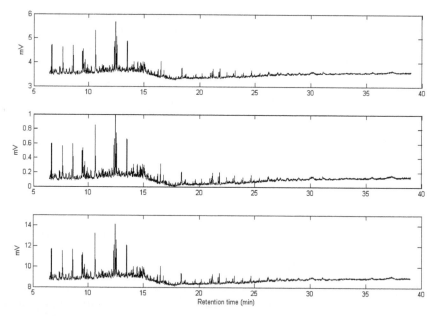

Fig. 6. Plots of original chromatogram (upper plot), normalization scaling chromatogram (middle plot) and Pareto scaling chromatogram (bottom plot).

Applications of PCA to the Monitoring of Hydrocarbon Content in Marine Sediments by Means of Gas
Chromatographic Measurements

91

3. Results and discussion

3.1 Application of PCA to hydrocarbon analysis in sediments from two areas of Italian coasts

Figure 7 reports the score plot of the first vs. the second factor obtained by PCA application to the GC chromatograms of superficial sediment samples taken along the coasts of Adriatic and Tyrrhenian sea. These two factors extracted by PCA explain the 90.7 % of a total variance of the chromatographic data set. This very high fraction of information, retained in two factors only, is an impressive example of PCA ability as "data reduction form"; now, the visual comparison of GC samples is possible by means of a simple two-dimensional plot depending on the reduction of the starting 20.000 variables (i.e. the retention times of hydrocarbons) to the two PCA factors.

The clustering of samples determining homogeneity and heterogeneity among samples is also evident and does not require further multivariate methods such ad discriminant analysis to investigate the classification of samples. Though these samples come from different seas and areas, some samples of the two areas have comparable hydrocarbon compositions as results from several VL and BG samples present in a same cluster, while samples of the Bagnoli area show different hydrocarbon compositions. This result means that the contributions of several biogenic (i.e. natural) and anthropogenic hydrocarbons can make sometimes comparable even sediments from different areas such as the two seas. These results can be hardly retrieved by the visual examination of the 29 chromatographic plots.

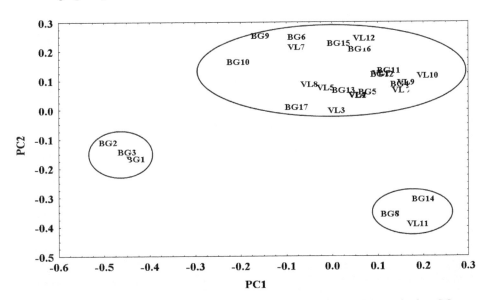

Fig. 7. Score plot of the first (PC1) vs. the second (PC2) factor from PCA applied to GC chromatograms of the Venice lagoon in Northern Adriatic (VL) and Bagnoli (BG) in Tyrrhenian sea. The two factors explain the 83.9 % and the 6.8 % respectively of the total variance. The ellipses are arbitrary and show the three different clusters.

However PCA can give additional information concerning the qualitative composition of samples because loading analysis can detect the hydrocarbon characteristics determining the similarities and dissimilarities observed in Figure 7.

The loading plot of the first factor (Figure 8) shows the generally high variability present in the hydrocarbon distribution of environmental samples as this factor explains the 83.9% of the total variance. Moreover, Figure 8 shows allows to retrieve characteristics concerning the hydrocarbon distribution of these samples. Pristane and phytane are two peculiar hydrocarbons able to characterise the biogenic and the anthropogenic sources present in environmental samples. In fact, pristane is a hydrocarbon typical of biogenic sources whereas phytane is a hydrocarbon typical of anthropogenic sources (Wang et al., 1999; Mecozzi et al., 2008; Duan et al., 2010). In this loading plot, pristane is negligible (retention time 15.5 minutes) whereas phytane is present (retention time 16.2 minutes in Figure 8, upper plot). In addition, the wax hydrocarbons (i.e. number of carbon higher than 24) which are also typical of biogenic sources (Wang et al., 1999; Duane et al., 2010; Ibbotson and Ibhadon, 2010; Ahad et al., 2011), are absent as shown by the negligible presence of chromatographic peaks with retention time higher than 20 minutes (Mecozzi et al., 2011). So the first loading plot describes the anthropogenic feature of the examined samples.

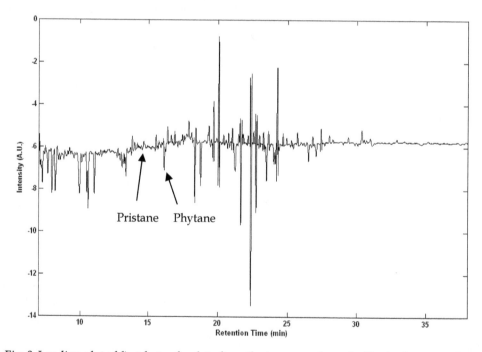

Fig. 8. Loading plot of first factor for data from the two areas from the Venice lagoon and Bagnoli near Naples. The arrows show the position of pristane (negligible) and phytane (present) in the corresponding chromatogram plot.

Applications of PCA to the Monitoring of Hydrocarbon Content in Marine Sediments by Means of Gas
Chromatographic Measurements

93

The loading plot of the second factor (Figure 9) , though explaining about the 7% of the total variance only, shows that samples in the upper cluster of Figure 7 are characterised by little concentration changes of some specific hydrocarbons related to biogenic hydrocarbon sources. In fact, with respect to the loading plot of Figure 8, here several linear hydrocarbons with carbon number higher than 24 are present and this is a marker of biogenic hydrocarbons (Wang et al, 1999; Duane et al., 2010). Obviously, due to the heterogeneity of the hydrocarbon composition, PCA can not specify the concentration changes of a single hydrocarbon, but in any case, it is relevant that we can compare samples of different origins solving the problem related to the general lack of methods to compare regional differences in areas submitted to potential hydrocarbon spills (Fraser at al., 2008).

Another interesting and useful advantage of using PCA in GC monitoring data consists of its support to the application of another well diffused unsupervised pattern recognition method such as Cluster Analysis. According to its name, CA performs the classification of data by identifying clusters of data having relevant similarities and for this purposes, it uses the multivariate distance among samples (Massart and Kaufmann, 1983).

CA is considered a fast screening method to perform exploratory data analysis though it does not identify the variables which determine similarity and or dissimilarity among samples; this remains a peculiar ability of PCA (Figures 8 and 9).

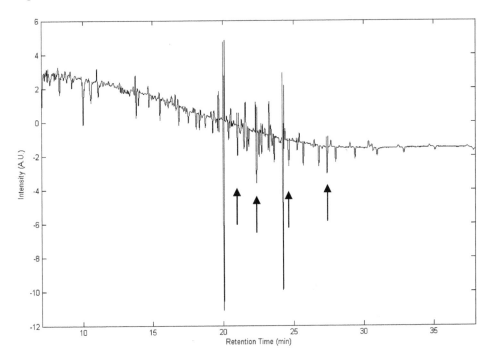

Fig. 9. Loading plot of second factor for data from the two areas from the Venice lagoon and Bagnoli near Naples. The arrows show the presence of some linear high molecular weight hydrocarbons, with more than 24 carbon atoms, typical of biogenic sources.

However, the application of CA to samples with over than 20.000 variables such as the case of the GC data is almost impossible due to computational and collinearity problems among variables (Massart and Kaufman, 1983). Conversely when CA is applied by means of the PCA scores, we have many peculiar advantages because this approach requires a small number of uncorrelated factors only while it does not require the use of specific distance such as the Mahalanobis one (Massart and Kaufman, 1983). This approach reported in Figure 10, shows that samples are clustered in a perfect agreement with Figure 7 obviously and now, by means of the data reduction of PCA, we can apply CA for estimating the percent of similarity existing among samples.

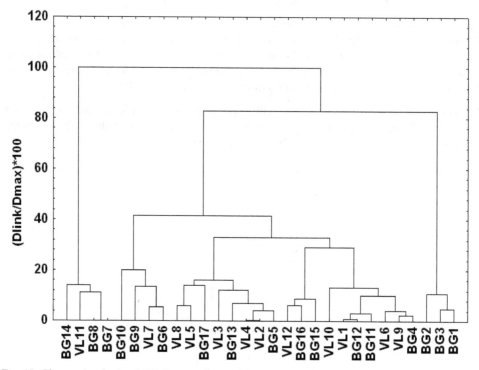

Fig. 10. Cluster Analysis of GC data performed by means of PCA scores. The ratio (Dlink/Dmax)* 100 of the ordinate axis is the quantitative measurements of the dissimilarity among samples.

3.2 Application of PCA to hydrocarbon analysis of sediment samples from an Antarctic core

The application of PCA to the chromatographic data set of an Antarctic sediment core (Figure 11) gives even more peculiar results with respect to those obtained in the previous section. Being Antarctic continent uncontaminated, we can suppose reasonably that hydrocarbons present in sediment core samples depend on biogenic contribution essentially with negligible anthropogenic contributions. If so, the hydrocarbon composition changes observed along the sections of the Antarctic sediment core should have a qualitative

Applications of PCA to the Monitoring of Hydrocarbon Content in Marine Sediments by Means of Gas
Chromatographic Measurements

95

homogeneous composition depending on the biogenic contributions. As a consequence, the observed quantitative changes should depend on the natural stratification events only. The results reported in the score plot of Figure 11 supports this hypothesis.

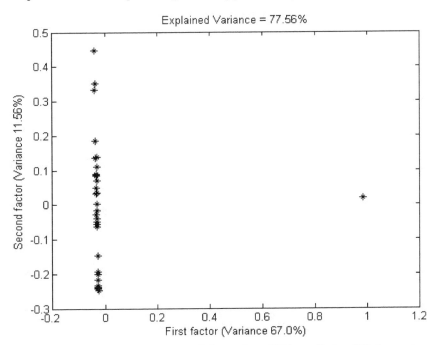

Fig. 11. Score plot of the first vs. the second factor from PCA applied to GC chromatograms of the Antarctic sediment core. The two factors explain the 67.0 % and the 11.56 % respectively of the total variance.

The first factors explains the 67.0 % of the total variance and the score values have an almost constant value while the positions of the samples changes with the scores of the second factor explaining the 11.56 % of the variance. The loading analysis reported in Figure 11, gives many details for the clarification of these findings. In the first factor, it is evident the presence of a significant hydrocarbon peak at high molecular weight (retention time close to 35 minutes) assigned to the linear hydriacrbon with 38 carbon atom number. This is a wax hydrocarbon, typical of biogenic contributions arising from the degradation of living cells (Duane et al., 2010).

PCA confirms the supposed prevalence of biogenic contributions for these samples depending on prevalent presence of the biogenic linear hydrocarbon with 38 carbon number, suggesting a significant homogeneous composition mostly governed by the natural stratification of sediments as well. In addition, if the hydrocarbon distribution along the sections core is determined by the natural stratification of sediments only, we can suppose that it is governed by time. In this case, we could test the hypothesis of the time depending relationship between stratification of hydrocarbon distribution in sediments by means of the autocorrelation function, a typical approach for time series analysis (Brereton, 2003). In fact, autocorrelation is a tool for studying time trend and periodicity present in an univariate data set according to the regressive model

$$Y_{t+1} = mY_t + \text{cost} \qquad t = 1, 2,\ldots\ldots n \qquad (6)$$

Autocorrelation has an easy application to univariate time series data but its application to multivariate data such chromatographic ones can be performed after a PCA data reduction, under the condition that its first factor explains a high percent of the total variance (Brereton, 2003). In the case of the Antarctic core samples this condition is fulfilled (i.e. 67% in the first factor) and the first factors can be considered as an univariate time series. So we can examine our data by the autocorrelation method using the score values of the first factor.

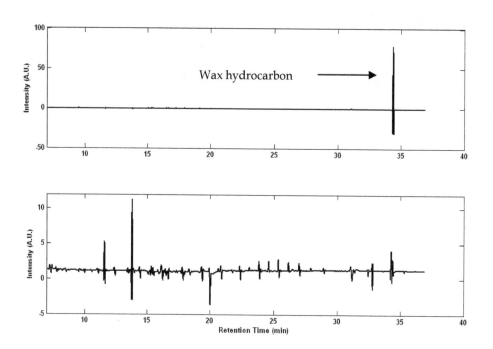

Fig. 12. Loading plot of the first (upper) and second (bottom) factors related to data of the Antarctic sediment core.

We report the result of this time series PCA application in Figure 13. The time descending trend of hydrocarbon distribution, depending on the natural stratification of sediments only, is clearly supported by the shape of the autocorrelation plot. On the base of this finding, we can verify that the hydrocarbon distribution shows a time trend depending on its biogenic contributions, because if anthropogenic sources were also present, we should observe a more irregular vertical profile and not the time trend supposed by Figure 11 and clearly confirmed by Figure 13.

Applications of PCA to the Monitoring of Hydrocarbon Content in Marine Sediments by Means of Gas
Chromatographic Measurements

97

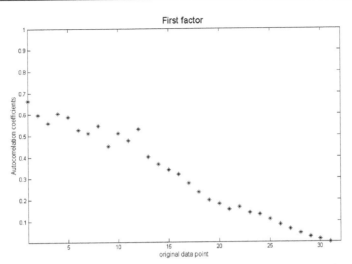

Fig. 13. Autocorrelation plot of the first factor of PCA applied to the hydrocarbon data
distribution. The abscissa axis corresponds to the number of the sediment sections of the
Antarctic core, while the number on the ordinate axis corresponds to the value of the
autocorrelation function determined by the first PCA factor.

4. Conclusion

In this study, we have presented all the aspects of pretreatment, scaling and use related to
the application of PCA to complex multivariate chromatographic data set coming from
environmental studies. As far as data pretreatment concerns, we stress the importance of
data redundancy reduction, signal to noise improvement and data scaling. For this latter
aspect, we have evidenced the peculiar advantages given by normalization and Pareto
scaling techniques with respect to the most applied autoscaling technique. When applied to
two specific cases of environmental studies PCA allows to retrieve much more information
than that obtained by the conventional visual examination of GC chromatograms. Both case
of studies show the power of PCA for explorative data analysis in chromatography and in
addition, its ability as "data reduction form "supports the use of other statistic and
complimentary techniques such CA and Time Series Analysis in the interpretation and
verification of the environmental results.

5. Acknowledgments

This study has been jointly supported by the research project "ASTRA" financed by the Ente
Nazionale Idrocarburi (ENI), Milan, Italy and by the National Research Antarctic Program
(PROGDEF09_125) financed by the Italian Minster of the University and Research.

6. Appendix

6.1 MATLAB routine for performing reduction of data redundancy, smoothing and baseline correction and removal

function [d,g]=gcpretreatment(chromatogram,factored);

```
% Routine for data redundancy reduction, Savitzky Goaly filtering and  baseline removal in
chromatograms
% Ref. "Practical Aspects of Chemometrics for Oil Spill Fingerprinting"
% J.H. Chrinstensen and G. Tomasi, J. Chromatography A, 1169 (2007) 1-122
% chromatogram  is  a data file with X (time),Y(mV) in ASCII format
% factored is a number specifying the entity of the data reduction
% for N  number of X,Y couples of signals, factored=2 gives  a final an N/2 length data file
% "g" is the reduced chromatogram
% "d" is the removed baseline
% "g" has to be saved  as ASCII file for further elaboration by PCA
% Data redundancy reduction
a=chromatogram(:,1);
b=chromatogram(:,2);
a(1:factored:length(a))=[];
b(1:factored:length(b))=[];
% Formation of reduced chromatogram (matrix)
c=[a,b];
%
% Baseline removal
rid=c(:,2);
% determination of baseline "d"
for i=1:length(rid)-1
    if i==1
    d(1)=rid(1);
    end
d(i)=rid(i+1)-rid(i);
end
d=[d(1),d];
f=rid-d';
%
% Savitzki Golay  filtering (third order function 17 points)
g=sgolayfilt(f,3,35);
subplot(3,1,1)
plot(chromatogram(:,1), chromatogram(:,2),'k')
title('Original Chromatogram')
xlabel('Time (min)')
ylabel('Intensity (mV)')
axis([min(chromatogram(:,1))  max(chromatogram(:,1))  min(g)  max(g)])
subplot(3,1,2)
plot(a,g,'k')
title('Chromatogram After Baseline Removal and Savitzky Golay Filtering')
xlabel('Time (min)')
ylabel('Intensity (mV)')
axis([min(a) max(a) min(g) max(g)])
subplot(3,1,3)
plot(a,d,'k')
title(' Removed Baseline ')
xlabel('Time (min)')
ylabel('Intensity (mV)')
```

axis([min(a) max(a) min(d) max(d)])

6.2 MATLAB routine for performing normalisation scaling

```
function [b]=norma(dataset);
% Routine for normalization of a spectral or chromatographic data samples
% Data matrix is column wise; each column corresponds to one sample
% dataset is the ASCII files of data to be normalized
% Files are uploaded as ASCII file
[m,n]=size(dataset);
for j=1:m
for i=1:n
b(j,i)=(dataset(j,i)-min(dataset(:,i)))/(max(dataset(:,i))-min(dataset(:,i)));
end
end
```

6.3 MATLAB routine for performing Pareto scaling

```
function [b]=paretoscaling(dataset);
% Routine for scaling of a spectral or chromatographic data samples
% by the Pareto approach
% Data matrix is column wise; each column corresponds to one sample
% dataset is the ASCII files of data to be normalized
[m,n]=size(dataset);
for j=1:m
for i=1:n
b(j,i)=(dataset(j,i))/sqrt(std(dataset(:,i)));
end
end
```

6.4 MATLAB routine for performing PCA

```
function [scores,loadings,varpercent]=pcawp(x);
% Principal Component Analysis (PCA)according to the algorithm Singular Value
% described by P. Geladi in Calculating Principal Component Loadings and Scores
% ISBN 91-7191-083-2, Umea, Sweden
% the "x" file is the data file after all the pretreatments
% Matrix r*c whit "r" rows (samples) and "c" columns (variables) can be analysed
% use the "save filename.txt a –ascii -tabs" instruction for saving files
% varpecent is the file with percent of variance explained by all the factors
[u,d,v]=svd(x);
% Determination of explained variance retained by each factor
l=d.*d;
varpercent=diag(l./trace(l))*100
scores=u*d;
loadings=v;
```

7. References

Ahad, J.M.E.; Ganeshram, R. S.; Bryant, C. L.; Cisneros-Dozal, L.; Ascough, P. L.; Fallick, A.E. & Slater, G.F. (2011). Sources of n-alkanes in un urbanized estuary: Insight from molecular distributions and compound-specific stable and radiocarbon isotopes. Marine Chemistry, Vol. 126, No. 1, 239-249, ISSN 0304-4203

Brereton, R. (2003). Chemometrics, Data Analysis for the Laboratory and Chemical Plant, John Wiley & Sons, West Sussex Po, UK, ISBN 0-471-48977-8

Christensen, J. H. & Tomasi, G. (2007). Practical Aspects of Chemometrics for Oil Spill Fingerprint. Journal of Chromatography A, Vol. 1169, No. 1-2, 1-22, ISSN 0021-9673

Cicero, A.; Mecozzi, M.; Morlino, R. ; Pellegrini, D. & Veschetti E. (2001). Distribution of Chlorinated Organic Pollutants in Harbor Sediments of Livorno (Italy): A Multivariate Approach to Evaluate Dredging Sediments. Environmental Monitoring & Assessment, Vol. 71, No.3, 297-316, ISSN: 0167-6369

Conti, M. E. & Mecozzi, M. Multivariate Approaches in Biomonitoring Studies in 'Biological Monitoring: Theory and Application' , M.E. Conti Editor, (2008). Southampton, UK, WIT PRESS, ISBN: 978-1-84564-002-6

Duane, M. ; He, K. & Liu, X, (2010). Characteristics and Source Identification of Fine Particulate n-alkanes in Beijing, China. Journal of Environmental Sciences, Vol. 22, No. 7, 998-1005, ISSN 1001-0742

Fraser. G.S.; Ellis, J. & Hussain, L. (2008). An International comparison of Governmental Disclosure of Hydrocarbon Spills from Offshore Oil and Gas Installation. Marine Pollution Bulletin, , Vol. 56, No.1, 9-13, ISS 0025-326X

Geladi P. (2002). Calculating Principal Component Loadings and Scores in MATLAB, ISBN 91-7191-083-2, Umeå, Sweden

Kokaly, M.; Rihtarič, M. & Kreft, S. (2011). Commonly Applied Smoothing of IR Spectra Showed Unappropriate for the Identification of Plant Leaf Samples. Chemometrics and Intelligent Laboratory Systems, Vol. 108, No.2, 154-161, ISS 0169-7439

Ibbotson, J. & Ibhadon. A.O. (2010). Origin and analysis of aliphatic and cyclic hydrocarbons in northeast United Kingdom coastal marine sediment. Organic Geochemistry, Vol. 60, No. 10, 1136-1141, ISSN0025-326X

Massart, D. L. & Kaufmann, L. 1989. The Interpretation of Analytical Chemical Data by the use of Cluster Analysis. Robert E. Krieger Publishing Company, Malabar, Florida, USA, ISBN 0-89464-358-4

Mecozzi, M. & Tomassetti, P. (2007). Handling of a Large Data Set: Application of Time Series Analysis to Oceanographic Studies. International Journal of Environment & Health, Vol. 1, No.6, 347-359, ISSN 1743-4955

Mecozzi M.; Scarpiniti, M.; Ragosta, E.; Pietroletti, M. & Di Mento, R. (2008). Proposal for a deconvolution procedure for the gas chromatographic estimation of pristane and phytane in marine sediments. International Journal of Environment & Health, Vol.3, No.1, 126-138, ISSN 1743-4955

Mecozzi, M.; Pietroletti, M.; Mattiello, S.; Moscato, F. & Oteri, F. (2011). Proceedings of 17° International Symposium on Separation Sciences, Cluj, Romanian, 5-9 September. ISBN 978-973-133-981-8

Noda, I. (2008). Scaling Techniques to Enhance two-Dimensional Correlation Spectra. Journal of Molecular Structure, Vol. 883-884, No.1, 216-227, ISS 0022-2860/S

Wang, Z.; Fingas, M. & Page, D.S. (1999). Oil Spill Identification. Journal of Chromatography A, Vol. 843, No. 1-2, 369-411, ISS 0021-9673(99)00120-X

6

EM-Based Mixture Models
Applied to Video Event Detection

Alessandra Martins Coelho[1] and Vania Vieira Estrela[2]

[1]*Instituto Federal de Educacao, Ciencia e Tecnologia do Sudeste de Minas Gerais*
(IF SUDESTE MG), Rio Pomba, MG,
[2]*Departamento de Telecomunicacoes, Universidade Federal Fluminense (UFF),*
Niterói, RJ,
Brazil

1. Introduction

Surveillance system (SS) development requires hi-tech support to prevail over the shortcomings related to the massive quantity of visual information from SSs [Fuentes & Velastin (2001)]. Anything but reduced human monitoring became impossible by means of its physical and economic implications, and an advance towards an automated surveillance becomes the only way out.

When it comes to a computer vision system, automatic video event comprehension is a challenging task due to motion clutter, event understanding under complex scenes, multi-level semantic event inference, contextualization of events and views obtained from multiple cameras, unevenness of motion scales, shape changes, occlusions and object interactions among lots of other impairments. In recent years, state-of-the-art models for video event classification and recognition [Zhang et al. (2011), Yacoob et al. (1999)] include modeling events to discern context, detecting incidents with only one camera (Ma et al. (2009), Zhao et al. (2002), Zelnik-Manor et al. (2006)], low-level feature extraction and description, high-level semantic event classification and recognition. Even so, it is still very burdensome to recuperate or label a specific video part relying solely on its content.

Principal component analysis (PCA) has been widely known and used, but when combined with other techniques such as the expectation-maximization (EM) algorithm its computation becomes more efficient.

This chapter introduces advances associated to the concept of Probabilistic PCA (PPCA) analysis [Tipping et al., 1999)] by of video event understanding technologies. The PPCA model-based method results from the combination of a linear model and the EM algorithm in an iterative fashion in order to determine a principal subspace (PS). Thus, additional work may be needed to find precise principal eigenvectors of the data covariance matrix, with no rotational uncertainty.

Kernel principal component analysis (KPCA) is a nonlinear PCA extension that relies on the kernel trick. It has received immense consideration for its value in nonlinear feature mining and other applications. On the other hand, the main drawback of the standard KPCA is that the huge amount of computation required, and the space needed to store the kernel matrix. KPCA can be viewed as a primal space problem with samples created via incomplete Cholesky decomposition. Therefore, all the efficient PCA algorithms can be easily adapted into KPCA. Furthermore, KPCA can be extended to a mixture of local KPCA models by applying the mixture model to probabilistic PCA in the primal space. The theoretical analysis and experiments can shed light on the performance of the proposed methods in terms of computational efficiency and storage space, as well as recognition rate, especially when the number of data points n is large.

By considering KPCA from a probabilistic point of view with the help of the EM algorithm, the computational load can be alleviated, but there still exists a rotational ambiguity with the resulting algorithm implementation. To unravel this intricacy, a constrained EM algorithm for KPCA (and PCA) was formulated founded on a coupled probability model. This brings in advantages related to many factors such as the necessary precision of extracted components, the number of the separated smaller data sets (which is usually empirically set), and the data to be processed. As a generic methodology, another thread of speeding up kernel machine learning is to seek a low-rank approximation to the kernel matrix. Since, as noted by several researchers, the spectrum of the kernel matrix tends to decay rapidly, the low-rank approximation often achieves satisfactory precision.

This chapter also aims at looking closely to ways and metrics in order to evaluate these less intensive EM implementations of PCA and KPCA.

2. Different ways of computing PCA

PCA is based on statistical properties of vector representations. It is an important tool for image processing because it decorrelates the data and compacts information [Xu (1998), Rosipal & Girolami (2001)].

PCA has been used profusely in all forms of analysis, since it is a straightforward, nonparametric way of extracting important information from ambiguous data sets. It helps reducing an intricate data set to a lower dimensional one that too often expose an unknown and simplified structure.

This section introduces three ways of calculating principal components (PCs): (i) via explicit computation of the covariance matrix C_X or, equivalently, XX^T; (ii) by means of the singular value decomposition (SVD) of the original problem so that it can be replaced by the calculation of $C_X=Y^TY$ with $Y=(n-1)^{1/2}X^T$, which requires the determination of eigenvectors of a system with smaller dimension; and (iii) using the EM algorithm.

Because (i) and (ii) are related to the concept of covariance, they require square matrices (XX^T and Y^TY, respectively). In the third case, there is an underlying probabilistic interpretation of the problem.

2.1 First approach: Solving PCA using eigenvectors of the covariance matrix

By seeking another basis, which is a linear combination of the original basis, the data set can be better represented. Linearity simplifies the problem because it restricts the set of prospective bases, and handles the implicit postulation of continuity in a data set. Let X be a matrix representing the original data set Y, be another matrix related by a linear transformation P that represents a change of basis. X is the original recorded data set and Y is its new representation

$$PX = Y. \tag{1}$$

Geometrically, P is a rotation and a stretch which again transforms X into Y. The covariance matrix of X is

$$C_X = \frac{1}{(n-1)} XX^T \tag{2}$$

PCA also takes for granted that mean and variance are sufficient statistics are enough to depict a probability distribution. This happens to be the case with exponential distributions (Gaussian, Exponential, etc). Deviations from an exponential distribution could nullify this assumption. Diagonalizing a covariance matrix might not give acceptable results. This hypothesis guarantees that the SNR and the covariance matrix totally portray noise and redundancies. The following factors can corrupt data: noise, rotation and redundancy. A common noise metric is the signal-to-noise ratio (SNR), or a ratio of variances σ^2 as follows:

$$SNR = \frac{\sigma_{signal}^2}{\sigma_{noise}^2} \tag{3}$$

A high $SNR (\gg 1)$ indicates clean data, while a low SNR points to noisy data. Large variances have important dynamics which means that the data is supposed to have a high SNR. Thus, principal components (PCs) with larger associated variances correspond to interesting dynamics, while those with lower variances may characterize noise.

Returning to (1), X is an $m \times n$ matrix, where m is the number of measurement types and n is the number of samples. The goal is to find an orthonormal P such that

$$C_Y = \frac{1}{(n-1)} YY^T \tag{4}$$

is diagonal and rows of P are the PCs of X. Because lots of real world data are normally distributed, PCA usually provides a robust solution to small deviations from this assumption. Rewriting C_Y in terms of P yields

$$C_Y = \frac{1}{(n-1)} YY^T = \frac{1}{(n-1)} (PX)(PX)^T$$

$$= \frac{1}{(n-1)} PXX^T P^T = \frac{1}{(n-1)} P(XX^T)P^T \Rightarrow \tag{5}$$

$$C_Y = \frac{1}{(n-1)} PAP^T$$

where $A \equiv XX^T$ is symmetric with

$$A = EDE^T. \tag{6}$$

The eigenvectors of A are arranged as columns of E and D is a diagonal matrix. A has $r \leq m$ orthonormal eigenvectors where r is the rank of the matrix. The rank of A is less than m when A is degenerate or all data reside in a subspace of dimension $r \leq m$. Maintaining the constraint of orthogonality, this situation can be remediated by selecting $(m - r)$ further orthonormal vectors to complete E. These additional vectors do not influence the final solution because the variances associated with these directions are zero.

Each row p_i is an eigenvector of XX^T and form $P \equiv E^T$. Combining the previous equations results in

$$A = P^T \text{DP}. \tag{7}$$

Since $P^{-1} = P^T$, then C_Y becomes

$$
\begin{aligned}
C_Y &= \frac{1}{(n-1)} PAP^T = \frac{1}{(n-1)} P(P^T DP)P^T \\
&= \frac{1}{(n-1)}(PP^T)D(PP^T) = \frac{1}{(n-1)}(PP^{-1})D(PP^{-1}) \\
&= \frac{1}{(n-1)} D
\end{aligned}
\tag{8}
$$

In practice, computing PCA of a data set X requires subtracting off the mean of each measurement type and the calculation of the eigenvectors of XX^T.

2.2 A more general solution: SVD

PCA relates closely to singular value decomposition (SVD), but SVD is a more general method to deal with change of basis. Let X be an arbitrary $n \times m$ matrix and $X^T X$ be a symmetric square $n \times n$ matrix with rank r. $V = \{v_1, v_2, \dots, v_r, 0, \dots, 0\}$ is the set of orthonormal eigenvectors associated with eigenvalues $\Sigma = Diag\{\sigma_1, \sigma_2, \dots, \sigma_r, 0, \dots, 0\}$ for the symmetric matrix $X^T X$ such that

$$(X^T X)v_i = \lambda_i v_i \, , \tag{9}$$

where $\sigma_i \equiv \sqrt{\lambda_i}$ are positive real singular values and $U = \{u_1, u_2, \dots, u_r, 0, \dots, 0\}$ is the set of orthonormal vectors defined by $u_i = (1/\sigma_i)Xv_i$. V and U contain, respectively, $(m-r)$ and $(n-r)$ appended zeros and $\sigma_1 \geq \sigma_2 \geq \cdots \geq \sigma_r$ are the rank-ordered set of singular values. The matrix version of SVD is given by

$$XV = U\Sigma. \tag{10}$$

Because V is orthogonal, multiplying both sides of the expression above by $V^{-1} = V^T$ leads to the final form of the SVD:

$$X = U\Sigma V^T \, , \tag{11}$$

which states that any arbitrary matrix X can be converted to an orthogonal matrix, a diagonal matrix and another orthogonal matrix as follows:

$$X = U\Sigma V^T \Rightarrow U^T X = \Sigma V^T \Rightarrow U^T X = Z \, , \tag{12}$$

where $Z \equiv \Sigma V^T$. Note that U^T is a change of basis from X to Z. The fact that the orthonormal basis U^T transforms column vectors means that U^T is a basis that spans the columns of X. Bases that span the columns are termed the column space of X. If $Z \equiv U^T \Sigma$, then the rows of V^T (or the columns of V) are an orthonormal basis for transforming X^T into Z. Because of the transpose of X, it follows that V is an orthonormal basis spanning the row space of X.

Matrices V and U are $m \times m$ and $n \times n$ respectively. Σ is a matrix with a small amount of non-zero values along its diagonal. The SVD allows for creating a new $m \times n$ matrix Y as follows:

$$Y \equiv \frac{1}{\sqrt{n-1}} X^T, \tag{13}$$

where each column of Y has zero mean. The definition of Y becomes obvious by looking at $Y^T Y$:

$$Y^T Y = (\frac{1}{\sqrt{n-1}} X^T)^T (\frac{1}{\sqrt{n-1}} X^T) = \frac{1}{n-1} X^{TT} X^T = \frac{1}{n-1} XX^T = C_X, \tag{14}$$

hence, Y is an $n \times m$ and by construction $Y^T Y$ equals the covariance matrix of X. The PCs of X are the eigenvectors of C_X. Applying SVD to Y, the columns of matrix V contain the eigenvectors of $Y^T Y = C_X$. Therefore, the columns of V are the PCs of X.

V spans the row space of $\equiv \frac{1}{\sqrt{n-1}} X^T$. Therefore, V must also span the column space of $\frac{1}{\sqrt{n-1}} X$. We can conclude that finding the PCs amounts to finding an orthonormal basis that spans the column space of X. If the final goal is to find an orthonormal basis for the column space of X then we can calculate it directly without constructing Y. By symmetry the columns of U produced by the SVD of $\frac{1}{\sqrt{n-1}} X$ must also be the PCs.

One benefit of PCA is that we can examine the variances C_Y associated with the principal components. Often one finds that large variances associated with the first $k < m$ PCs, and then a precipitous drop-off. One can conclude that most interesting dynamics occur only in the first k dimensions.

Both the strength and weakness of PCA is that it is a non-parametric analysis. When data are not normally distributed PCA fails. In exponentially distributed data, the axes with the largest variance do not correspond to the underlying basis. There are no parameters to tweak and no coefficients to adjust based on user experience: the answer is unique and independent of the user.

This also poses a problem, if some system characteristics are not known *a-priori*, then it makes sense to incorporate these assumptions into a parametric algorithm or an algorithm with selected parameters.

This prior non-linear transformation is sometimes termed a kernel transformation and the entire parametric algorithm is called $KPCA$. This procedure is parametric because the user must incorporate prior knowledge of the structure in the selection of the kernel but it is also more optimal in the sense that the structure is more concisely described.

One might envision situations where the PCs need not be orthogonal. Only the subspace is unique because the PCs are not uniquely defined. In addition, eigenvectors beyond the rank of a matrix (i.e. $\sigma_i = 0$ for $i > rank$) can be selected almost capriciously. Nevertheless, these degrees of freedom do not influence the qualitative features of the solution nor a dimensional reduction.

For instance, if an image contains a 2-D exponentially distributed data set, then the largest variances will not correspond to the meaningful axes and PCA fails.

2.3 EM algorithm for PCA

There is a close relationship between the expectation-maximization (EM) algorithm and PCA, which leads to a faster implementation of the PPCA. The algorithm extracts a small number of eigenvectors and eigenvalues from large sets of high dimensional data. It is computationally efficient in space and time and does not require computing the sample covariance of the data.

PCA is largely used in data analysis due to its optimality in terms of mean squared error, and its linear scheme to reduce the dimensions of vectors, so that compression and decompression become simple operations to carry out given the model parameters. Notwithstanding these interesting features, PCA has some deficiencies. The other two methods for finding the PCs are impractical for high dimensional data. Difficulties can arise in both computational complexity and data scarcity when diagonalizing a covariance matrix of n vectors in a p-dimensional space when n and p amount to hundreds or several thousands of elements. It is often the case that there is not enough data in high dimensions for the sample covariance to be of full rank (data scarcity). Moreover, care needs to be taken in order to use techniques such as the SVD, which do not need full rank matrices. Complexity makes the direct diagonalization of a symmetric matrix with thousands of rows tremendously expensive (it is $O(p^3)$ for $p \times p$ inputs). There are procedures such as the one proposed by Wilkinson (1965) which is $O(p^2)$ that decrease this cost when only the first most important eigenvectors and eigenvalues are necessary. The sample covariance calculation calls for $O(np^2)$ operations.

In most cases, the explicit computation of the sample covariance matrix should be avoided. Methods such as the snap-shot algorithm from Sirovich (1987), which has complexity of $O(n^3)$, take for granted that the eigenvectors sought out are linear combinations of the data points. In this section, a version of the EM algorithm from Dempster (1977) is presented for learning the PCs of a dataset. The algorithm does not require computing the sample covariance and has a complexity limited by $O(knp)$ operations, where k is the number of leading eigenvectors to be learned.

Usual PCA approaches cannot handle missing values: incomplete data must either be discarded or completed via ad-hoc interpolation techniques. A possible and uncomplicated solution is to replace missing coordinates with the mean of the known values in the corresponding coordinate or with estimation values relying on the known values. The EM algorithm for PCA benefits from the estimation of the maximum likelihood (ML) values for missing information directly at each iteration as stated by Ghahramani & Jordan (1994).

As a final point, independently of the technique used to perform PCA, there is no accurate probability model in the input space, because the probability density is not normalized in the PS . This means that once applying PCA to some data, the only criterion on hand to verify if the new data fit well the model is the squared distance of the new data from their projections into the PS. A data point distant from the training data but close to the PS will have a high pseudo-likelihood or low error. This chapter also brings in a model called sensible PCA (SPCA) which delineates a proper covariance structure in the data space as proposed by Roweis (1998) whose main contribution was to alleviate the computational load of the other two techniques with the help of the EM algorithm.

PCA can be interpreted as a limiting case of a particular class of linear-Gaussian models (LGMs), because these models capture the covariance structure of an observed p-dimensional variable y using less than $p(p+1)/2$ free parameters when compared to the full covariance matrix calculation. LGMs do this by assuming that y is the result from a linear transformation of some k-dimensional x plus additive Gaussian noise. Denoting the transformation by the $p \times k$ matrix C and the (p-dimensional) noise by v (with covariance matrix R) the generative model can be written as

$$y = Cx + vx, \tag{15}$$

where $x \sim \mathcal{N}(0, I)$ and $v \sim \mathcal{N}(0, R)$. x is considered independent and identically distributed (iid) according to a unit variance spherical Gaussian. Since v is also iid and independent of x, the model reduces to a single Gaussian model for y as follows:

$$y \sim \mathcal{N}(0, CC^T + R) . \tag{16}$$

With the purpose of saving parameters over the direct covariance representation in p-space, it is indispensable to select $k < p$ and to curb the covariance structure of v by constraining R. The second constraint allows the model to capture any interesting or informative projections in x. If R was not limited, then the algorithm could choose $C = 0$ and R would be the sample covariance of the data considering any deviation in the data as noise.

There are two central problems of interest when working with LGMs. Firstly, the compression problem asks if given fixed model parameters C and R, it is possible to gather information about the unknown x given a few observations y must be gathered. Since the data points are independent, one needs the posterior probability $P(x|y)$ given the corresponding single observation, resulting in

$$P(x|y) = \frac{P(y|x)P(x)}{P(y)} = \mathcal{N}(\beta y, I - \beta C)|x, \tag{17}$$

where $\beta = C^T(CC^T + R)^{-1}$ gives the expected value βy of the unknown and an estimate of the uncertainty in this value in the form of the covariance $(I - \beta C)$. y from x can be obtained from $P(x|y)$. Finally, the likelihood of any data point y comes from (16).

The second problem is called learning or parameter fitting. It seeks the matrices C and R that assign the highest likelihood to the observed data. There is a family of EM algorithms employing the inference formula above in the E-step to estimate the unknown and then choose C as well as R in the M-step, in order to maximize the expected joint likelihood of the estimated x and the observed y.

PCA is a limiting case of the LGM as the covariance of the noise v becomes infinitesimally small and equal in all directions, that is $R = \lim_{\varepsilon \to 0} \varepsilon I$. This makes the likelihood of y subject exclusively to the squared distance between it and its reconstruction C_X. The directions of the columns of C which minimize this error are the PCs. Inference now becomes a simple least squares projection:

$$P(x|y) = \mathcal{N}(\beta y, I - \beta C)|x, \text{ with } \beta = \lim_{\varepsilon \to 0} C^T (CC^T + \varepsilon I)^{-1} \tag{18}$$

or alternatively,

$$P(x|y) = \mathcal{N}((C^T C)^{-1} C^T y, \ 0)|_x = \delta(x - (C^T C)^{-1} C^T y) \tag{19}$$

Given that the noise became insignificant, the posterior over x collapses to a single point and the covariance turns out to be zero. Albeit the PCs can be computed explicitly, there is still an EM algorithm for the limiting case of zero noise. It can be easily derived from the standard algorithms (Sangers (1989), Oja (1989), Everitt (1984), Ghahramani et al. (1997)) by replacing the common E-step by the above projection as follows:

$$\text{E-step: } X = (C^T C)^{-1} C^T Y \tag{20}$$

$$\text{M-step: } C = YX^T (XX^T)^{-1}, \tag{21}$$

where Y is a $p \times n$ matrix containing all the observed data and X is a $k \times n$ matrix with the unknowns. The columns of C span the space of the first k PCs. To explicitly compute the corresponding eigenvectors and eigenvalues, the data can be projected onto this k-dimensional subspace to construct an ordered orthogonal covariance basis. This means that once an orientation for the PS was guessed, the presumed subspace is corrected and the data y is projected onto it to give the values of x. Next, the values of x are corrected and a subspace orientation is chosen to minimize the squared reconstruction errors of the data points.

Bear in mind that if C is $p \times k$ with $p > k$ and is rank k then left multiplication by $C^T (CC^T)^{-1}$, which appears not to be well defined because (CC^T) is not invertible, is exactly equivalent to left multiplication by $(C^T C)^{-1} C^T$. This is the same as the SVD idea of defining the inverse of the diagonal singular value matrix as the inverse of an element except if it is zero when it stays zero. The perception is that even if CC^T in fact is not invertible, the directions along which it is not invertible are just those that C^T is about to project out.

The EM algorithm for PCA amounts to an iterative procedure for finding the k leading eigenvectors without explicit computation of the sample covariance. Its complexity is limited by $O(knp)$ per iteration and so depends only linearly on both the dimensionality of the data and the number of points. Explicitly computing the sample covariance matrix result in complexities of $O(np^2)$, while other methods that form linear combinations of the data must calculate and diagonalize a matrix with all possible inner products between points and as a result have $O(n^2 p)$ complexity.

According to Roweis (1998), the standard convergence proofs for EM given by Dempster (1977) are appropriate to this algorithm as well, so a solution will always attain a local maximum of likelihood. Additionally, it is assumed that PCA learning do not have a stable maxima other than the global optimum which results in convergence to the true PS. The rate of convergence depends on the ratio of the largest eigenvalue to the second largest eigenvalue; the closer the two are in magnitude the slower the convergence will be.

In the complete data setting, the values of the projections x are viewed as missing information for EM. During the E-step these values are computed by means of projecting the observed data into the current subspace. This minimizes the model error given the observed data and the model parameters. However, if some of the input points lack certain coordinate values, then those values can be easily estimated in the same fashion. The E-step can be generalized as follows:

E-step: For each (possibly incomplete) point y find the unique pair of points x^* and y^* (such that x^* lies in the current PS and y^* lies in the subspace defined by the known coordinates of y) which minimize the norm $\| Cx * -y^* \|$. Set the corresponding column of X to x^* and the corresponding column of Y to y^*.

If y is complete then $y^* = y$ and x^* is found exactly as before. If not, then x^* and y^* are the solution to a least squares problem and can be found by, for instance, QR factorization. Observe that this method is not restricted to missing coordinates in the data; the unknown degrees of freedom may lie in any directions in the space. This outperforms replacing each missing coordinate with the mean of known coordinates.

3. EM algorithm for Sensible PCA (SPCA)

If R must have the form εI, but do not take the limit as $\varepsilon \to 0$, then this model is called SPCA according to Roweis (1998). The columns of C are still known as the PCs. From now on, the scalar value ε on the diagonal of R is called global noise level. It is worth noting that SPCA uses $1 + pk - k(k-1)/2$ free parameters to model the covariance. Once again, inference is done with (17) and learning by an EM algorithm. Because it has a finite noise level, SPCA defines the following model and probability distribution in the data space:

$$y \sim \mathcal{N}(0, CC^T + \varepsilon I) \tag{22}$$

which makes possible to evaluate the actual likelihood of new test data under an SPCA model. Furthermore, this likelihood will be much lower for data far from the training set even if they are near the PS, unlike the reconstruction error from PCA. The EM algorithm for SPCA is:

$$\text{E-step: } \beta = C^T (CC^T + \varepsilon I)^{-1} \mu_x = \beta Y \ \Sigma_x = nI \ - n\beta C + \mu_x \mu_x^T \tag{23}$$

$$\text{M-step: } C = Y \mu_x^T \Sigma^{-1} \varepsilon \ = \text{trace} [XX^T - C\mu_x Y^T]/n^2 \tag{24}$$

Since εI is diagonal, the inversion in the E-step can be performed efficiently using the matrix inversion lemma:

$$(CC^T + \varepsilon I)^{-1} = (I/\varepsilon - C(I + C^T C/\varepsilon)^{-1} C^T/\varepsilon^2) . \tag{25}$$

Because only the trace of the matrix in the M-step is taken, there is no need to compute the full sample covariance XX^T. Instead only the variance along each coordinate need to be computed. These two observations suggest that for small k, learning for SPCA also have complexities limited by $O(knp)$ and not worse.

4. EM algorithm for KPCA

Tipping & Bishop (1999) analyzed PCA from a probabilistic point of view and realized that probabilistic PCA (PPCA) is a special case of factor analysis. In addition, Rubin & Thayer (1984) developed the expectation-maximization (EM) learning algorithm for factor analysis. So, considering PPCA within the factor analysis framework, the principal components can be straightforwardly extracted by using the EM algorithm rather than performing eigenvalue decomposition. So, the computational burden on high dimensional data can be alleviated. Rosipal and Girolami (2001), transformed the EM procedure from data space to a nonlinearly related feature space. Thus, an EM approach to kernel PCA (KPCA) has arisen, which is very useful to find the nonlinear PCs. Scholkopf et al. (1998) introduced KPCA and demonstrated its value in machine learning as well as pattern recognition.

- KPCA needs to diagonalize the kernel matrix K, whose dimensionality N is equal to the number of data points and as the data set increases, KPCA becomes less viable due to the augmenting computational complexity $O(N^3)$, which prohibits it from being used in many applications. Moreover, there is still the problem of numerical precision when diagonalizing large matrices directly according to Rosipal and Girolami, (2001). So, the EM approach to KPCA (with computational complexity $O(qN^2)$ per iteration, where q is the number of extracted components) is a good remedy. Still, there exists a rotational ambiguity with the EM algorithm for PCA (and KPCA), which is unwanted from a theoretical point of view.

- Ahn & Oh (2003) have introduced a constrained EM algorithm by using a coupled latent variables model. Their proposed EM approach can directly compute the eigen-system of sample covariance matrix in data space as well as that of the kernel matrix. For the most part, when it is applied to the kernel matrix K, it is a dual form of the constrained EM algorithm for performing KPCA.

4.1 EM algorithm for any positive semi-definite matrix

Let $Y = [y_1, ..., y_N]$ be the matrix consisting of N p-dimensional vectors known as observations, and $X = [x_1, ..., x_N]$ be the q-dimensional latent variables associated with the data points of Y. The linear model relating an observed data vector y_n to a corresponding latent variable x_n is given by

$$y_n = W^T x_n + \varepsilon_n ,\qquad(26)$$

with $n = 1, ..., N$; the parametrical matrix $W \in \mathbb{R}^{p \times q}$ determines the connection between the data space, and the latent space. The p- dimensional noise vector ε_n is normally distributed with zero mean and covariance matrix $\sigma^2 I$. Vector x_n is also zero mean and normally distributed with identity covariance. By marginalizing with respect to x_n and optimizing W using the ML principle, Tipping & Bishop (1999) proved that the ML solution correspond to the situation when W spans the PS of the observed data (PPCA model).

The EM approach to PCA is a least-squares projection, which - as said by Rosipal & Girolami (2001), Xu (1998) and Tipping & Bishop (1999) - is given by:

$$\text{E-step: } X = (W^T W)^{-1} W^T Y, \text{ and}\qquad(27)$$

$$\text{M-step: } = Y X^T (X X^T)^{-1} .\qquad(28)$$

Later, by means of a coupled latent variables model along with PPCA, Ahn & Oh (2003) introduced a constrained EM algorithm for PCA:

$$\text{E-step: } X = \{\mathcal{L}(W^T W)\}^{-1} W^T Y, \text{ and} \tag{29}$$

$$\text{M-step: } = Y X^T \{\mathcal{U}(X X^T)\}^{-1}, \tag{30}$$

where the element-wise lower operator $\mathcal{L}$ was defined such that $\mathcal{L}(a_{ij}) = a_{ij}$, for $i \geq j$ and zero otherwise, and the upper operator $\mathcal{U}$ corresponds to $\mathcal{U}(a_{ij}) = a_{ij}$, for $i \leq j$ and zero if not. Ahn & Oh (2003) verified that as the noise level became infinitesimal, W would converge to be the ML estimator $W = U \Lambda^{1/2}$, where the q columns of U were the eigenvectors of the sample covariance matrix $(1/N) Y Y^T$ and Λ contained the related eigenvalues arranged diagonally in descending order of magnitudes. This removes the rotational ambiguity of algorithm as stated in (27) and (28).

The EM algorithm for PCA considers the latent variables $\{x_n\}$ as missing data. The E-step of the LM algorithm evaluates the expectation of the corresponding complete-data log-likelihood with respect to the posterior distribution of x_n given the observed y_n. The expectations $E(x_n|y_n)$ and $E(x_n x_n^T|y_n)$ form the basis of the E-step. In the M-step, the parameter W is updated to maximize the expected complete-data log-likelihood function, which is guaranteed to increase the likelihood of the observed samples $\{y_n\}$ as follows:

$$W = \left(\sum_{n=1}^N y_n E(x_n|y_n)^T\right)\left(\sum_{n=1}^N y_n E(x_n x_n^T|y_n)\right). \tag{31}$$

Combining the E-step and M-stes yields

$$W = Y Y^T W (W^T W)^{-1}((W^T W)^{-1} W^T Y Y^T W (W^T W)^{-1})^{-1} \tag{32}$$

as the noise level becomes infinitesimal.

Now from (32), if]et $S = Y Y^T$ and use the co upled latent variables model Ahn & Oh (2003), then the EM algorithm for PCA could be rewritten as

$$\text{E-step: } Z = \{\mathcal{L}(W^T W)\}^{-1} W^T, \text{ and} \tag{33}$$

$$\text{M-step: } W = S Z^T \{\mathcal{U}(Z S Z^T)\}^{-1}. \tag{34}$$

The modified constrained EM algorithm comes from (33) and (34), can be further generalized to any positive semi-definite matrix S. In fact, by the incomplete Cholesky decomposition any positive semi-definite matrix can be factorized as

$$S = L L^T, \tag{35}$$

where $L \in \mathbb{R}^{p \times r}$ and S have rank r. The columns of L are samples of y_n and apply the model (26). The modified constrained EM algorithm comes from (33) and (34). After convergence, the normalized columns of W are the leading q eigenvectors of S (please, refer to Roweis (1998) and Dempster (1977)). The related eigenvalues are the diagonal elements of $W^T S W$. The computational complexity of the proposed EM algorithm is $O(q p^2)$ per iteration.

The modified constrained EM algorithm (MCEM) even though simple in the derivation, is very useful in computing not only the eigen-system of sample covariance matrix in data space, but also that of kernel-based nonlinear algorithms. For instancee, the MCEM can be applied to KPCA directly. Given the set of N observations Y, the basic idea is to first map these input data into some new feature space via a nonlinear function ϕ, followed by standard linear PCA using the mapped samples $\phi(y_i)$. Thus, KPCA turns out to compute the most important eigenvectors of the $N \times N$ kernel matrix K which is defined such that the elements

$$K_{ij} = k(y_i, y_j) = (\phi(y_i) \cdot \phi(y_j)), \tag{36}$$

where k is the kernel function which calculates the dot product between two mapped samples $\phi(y_i)$ and $\phi(y_j)$. Hence, the mapping of ϕ does not need to be computed explicitly. If k is a positive definite kernel, then there exists a mapping ϕ into the dot product space G such that (36) holds.

To compute the leading eigenvectors of K, it is enough to replace S with K using the MCEM algorithm which can be viewed as a dual form of the constrained EM algorithm for performing KPCA. The computational complexity is $O(qN^2)$ per iteration. The projection of a test point y, whose image is $\phi(y)$ onto the q nonlinear principal axes is given by $W^T K_y$, where the columns of W are normalized, i.e., the i-th column w_i is divided by $\sqrt{w_i^T K w_i}$, such that the eigenvectors of the sample covariance matrix have unitary norms, and K_y is the vector $(k(y_1 \cdot y), \ldots, k(y_n \cdot y))^T$ (please, see Xu (1998)).

5. PCA in video event detection

Visual surveillance demands video sequence understanding, the detection of predefined events prone to activate an alarm, the tasks to performed, environment/scenario acquaintance and, consequently, a superior computational performance [Siebel and Maybank (2002), Fuentes and Velastin (2001)]. Nevertheless, smart SSs face the intricate task of analyzing people and their activities. A number of clues may be acquired from the investigation of people trajectories and their relations (tracking). The investigation of a single blob location or path can decide whether a person is located in an illegal area, running, jumping or hiding as in Figures (1) and (2). These data from two or more people may reveal facts regarding their interaction.

The amount of people present in a scene is called density. Events may also be classified into as position-based and dynamic-based events. SSs need to handle more than one image channel or cameras simultaneously in real time to be effectively applied to security, and this calls for a simplification of the image processing stages. A background recognition technique relying on motion detection along with a simple tracking algorithm to extract real-time blob and scene features such as blob position, blob speed and people density can help building semantic descriptions of predefined occurrences. Contrasting sequence parameters with the semantic description of the associated events related to the current scenario, the system is able to spot them and to alert the ultimate decision-maker. The case study presented here is concerned with people-oriented SSs applied to public environments.

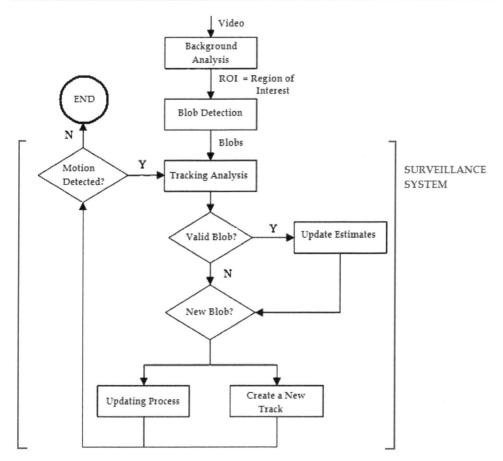

Fig. 1. Surveillance system using blob detection.

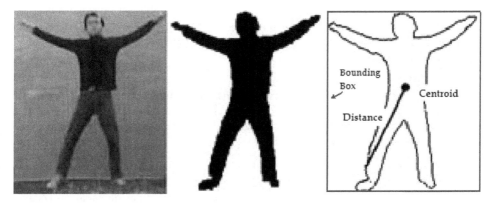

Fig. 2. Blob representation by means of a bounding box.

Background regions can be segmented out by detecting the presence or absence of motion between at least two consecutive frames as can be seen in Figure (3). An SS may hint that an event has indeed occurred because of frame-by-frame blob investigation and tracking, which offer sufficient data to analyze some prior events and the occurrence odds of others in a video sequence.

frame 1   frame 2   1. background   2. phone regions

3. head region   4. background   5. phone regions   6. head region

Fig. 3. Background and ROIs for the Susie sequence [Zhang et al. (2011)].

A matching procedure involving blobs from the current and preceding frames, and the overlapping of bounding boxes [Rivera et al. (2004)] can be used as evaluation criteria for ruling the interaction among people and things. This condition has proved helpful in other techniques [MacKenna et al. (2000)] and it does not necessitate the calculation of the blob position since blob motions are always considered smaller than their dimensions. Each new blob can be updated by means of the data accumulated from previous frames. If a new blob emerges, then the centroid position can be used to construct a new path. If two blobs join to form a new one, then this blob is labelled as a group and the information about combined blobs is stored for future use. This new blob group can be tracked independently. If the group splits once more, the system uses motion direction and blob characteristics to properly classify splitting blobs. Trajectories of single persons or cars can be effortlessly obtained following centroids in adjacent frames throughout a video sequence. Whenever needed, the tracked blob position can be interpolated before and after it became part of a group.

Camera networks, control rooms along with human resources and operators for surveillance have prompted lots of interest in automation of inspection tasks. These systems are also concerned with citizens' safety, people flow patterns (for counting purposes or as an aid to facilities planning), overcrowding of restricted or semi-open areas, atypical crowd movements, obstruction of exits, brawls, vandalism, falls/accidents, unattended objects, invasion of forbidden regions, and so on [Fuentes (2002)].

Blob recognition offers $2D$ information about people coordinates in a $3D$ setting. A more accurate position calls for either geometric camera calibration or stereo processing. Some examples of event detection that can be done by means of centroid position analysis [Zhang et al. (2011), Fuentes et al. (2002)] are

1. Unattended objects laying on the floor (blobs are normally smaller than people) presenting little or no motion, people falling on the ground and objects that move away from a person or reference point can be detected by means of a time analysis.
2. Suspicious activities related to hidden people may correspond to blobs disappearance along several consecutive frames.
3. Vandalism may involve isolation of one person/group present in the scene with irregular centroids motion and, possibly, changes in the background.
4. Temporal thresholding may be used to detect invasions and activate the alarms in case necessary actions must be taken.
5. Prevention of Attacks and Fights: Using people sensation of distance, and knowledge on social patterns of conduct, it is possible to establish a range of distances and profiles corresponding to different types of social interaction [MacKenna et al. (2000)]. By means of the analysis of fast centroids changes, important clues about the people present in a scene can be gathered, such as blob coincidence, merging of blobs and blob splitting to name a few.

Another way of posing the blob detection algorithm is to cluster displacement vectors and, then learn the blob centroids motions.

5.1 Case study: Motion estimation

Motion provides important information. Significant events, such as collision paths, object docking, sensor obstruction, object properties and occlusion can be characterized and understood with the help of the optic flow (OF). Segmenting an OF field (OFF) into coherent motion groups and estimating each underlying motion are very challenging tasks when a scene has several independently moving objects. The problem is further complicated by noise and/or data scarcity.

The main problem with motion analysis is the difficulty to get accurate motion estimates without prior motion segmentation and vice-versa. Pel-recursive (PR) schemes [Franz & Krapp (2000), Franz & Chahl (2003), Kim et al. (2005), Tekalp (1995)] can theoretically overcome some of the limitations associated with blocks by assigning a unique motion vector to each pixel.

Segmenting OF via EM algorithm for mixtures of PCs can be done successfully [Estrela & Galatsanos (2000), Tipping & Bishop (1999)] because both techniques share a close relationship. Most methods assume that there is little or no interference between the individual sample constituents or that all the constituents in the samples are known ahead of time. In real world samples, it is very unusual, if not entirely impossible, to know the entire composition of a mixture sample. Sometimes, only the quantities of a few constituents in very complex mixtures of multiple constituents are of interest [Blekas et al. (2005), Kim et al. (2005), Tipping & Bishop (1999)]. This section intends to solve OF problems by means of two different takes on PCA regression (PCR): 1) a combination of regularized least squares (RLS) and PCA (PCR_1); and 2) RLS followed by regularized PCA regression (PCR_2). Both involve simpler computational procedures than previous attempts at addressing mixtures [Blekas et al. (2005), Kim et al. (2005), Tipping & Bishop (1999), Jolliffe (2002), Wold et. al. (1983)].

5.1.1 Problem formulation

The displacement of every pixel in each frame forms the displacement vector field (DVF) and its estimation can be done using at least two successive frames. The goal is to find the corresponding intensity value $I_k(r)$ of the k-th frame at location $r = [x, y]^T$, and $d(r) = [d_x, d_y]^T$ the corresponding displacement vector (DV) at the working point r in the current frame by means of algorithms that minimize the DFD function in a small area containing the working point assuming constant image intensity along the motion trajectory. The perfect registration of frames will result in $I_k(r)=I_{k-1}(r-d(r))$ as seen in Figure (4). Figure (5) shows some examples of pixel neighborhoods. The DFD represents the error due to the nonlinear temporal prediction of the intensity field through the DV and is given by

$$\Delta(r;d(r))=I_k(r)-I_{k-1}(r-d(r)) .$$

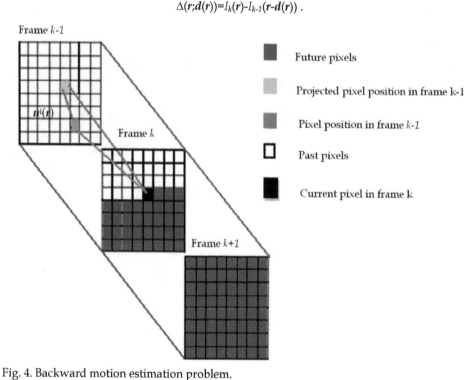

Fig. 4. Backward motion estimation problem.

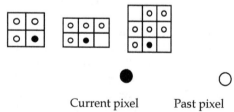

Fig. 5. Examples of causal masks.

An estimate of $\mathbf{d}(\mathbf{r})$, is obtained by directly minimizing $\nabla(r,d(r))$ or by determining a linear relationship between these two variables through some model. This is accomplished by using a Taylor series expansion of $I_{k-1}(r-d(r))$ about the location $(r-d^i(r))$, where $d^i(r)$ represents a prediction of $d(r)$ in the i-th step. This results in $\Delta(\mathbf{r},\mathbf{r}\text{-}\mathbf{d}^i(\mathbf{r})) = -\mathbf{u}^T \nabla I_{K-1}(\mathbf{r}\text{-}\mathbf{d}^i(\mathbf{r})) + e(\mathbf{r},d(\mathbf{r}))$, where the displacement update vector is $\mathbf{u}=[u_x, u_y]^T = d(r) - d^i(r)$, $e(r, d(r))$ stands for the truncation error resulting from higher order terms (linearization error) and $\nabla=[\partial/\partial_x, \partial/\partial_y]^T$ represents the spatial gradient operator. Considering all points in a neighborhood of pixels around r gives

$$\mathbf{z} = \mathbf{Gu} + \mathbf{n} \text{ ,}$$

where the temporal gradients ∇ $(r, r\text{-}d^i(r))$ have been stacked to form the $N\times1$ observation vector $\mathbf{z}$ containing DFD information on all the pixels in a neighborhood, the $N\times2$ matrix $\mathbf{G}$ is obtained by stacking the spatial gradient operators at each observation, and the error terms have formed the $N\times1$ noise vector n. The PR estimator for each pixel located at position $\mathbf{r}$ of a frame k can be written as

$$d^{i+1}(r) = d^i(r) + u^i(r),$$

where $u^i(r)$ is the current motion update vector obtained through a motion estimation procedure that attempts to find u, $d^i(r)$ is the DV at iteration i and $\mathbf{d}^{i+1}(\mathbf{r})$ is the corrected DV. The ordinary least squares (OLS) estimate of the update vector is

$$\mathbf{u}_{LS} = (\mathbf{G}^T\mathbf{G})^{-1}\mathbf{G}^T\mathbf{z} \text{ ,}$$

which is given by the minimizer of the functional $J(u)=\|z\text{-}Gu\|^2$. The assumptions made about n for least squares estimation are $E(n) = 0$, and $\text{Var}(n) = E(nn^T) = \sigma^2 I_N$, where $E(n)$ is the expected value (mean) of $\mathbf{n}$, and I_N, is the identity matrix of order N. From now on, $\mathbf{G}$ will be analyzed as being an $N\times p$ matrix in order to make the whole theoretical discussion easier. Since $\mathbf{G}$ may be very often ill conditioned, the solution given by the previous expression will be usually unacceptable due to the noise amplification resulting from the calculation of the inverse matrix $\mathbf{G}^T\mathbf{G}$. In other words, the data are erroneous or noisy.

The regularized minimum norm solution also known as regularized least square (RLS) solution is given by

$$\hat{\mathbf{u}}_{RLS}(\Lambda) = (\mathbf{G}^T\mathbf{G}+\Lambda)^{-1}\mathbf{G}^T\mathbf{z} \text{ .}$$

The RLS estimate of the motion update vector can be improved by a strategy that uses local properties of the image. Each row of $\mathbf{G}$ has entries $[g_{xi}, g_{yi}]^T$, with $i = 1, ..., N$. The spatial gradients of I_{k-1} are calculated through a bilinear interpolation scheme similar to what is done in [Estrela & Galatsanos (2000), Estrela & Galatsanos (1998)]. The entries $f_{k-1}(\mathbf{r})$ corresponding to a given pixel location inside a causal mask is needed to compute the spatial gradients by means of bilinear interpolation [Estrela & Galatsanos (2000), Estrela & Galatsanos (1998)] at location $\mathbf{r} = [x,y]^T$ as follows:

$$\begin{bmatrix} \theta_x \\ \theta_y \end{bmatrix} = \begin{bmatrix} x - \lfloor x \rfloor \\ y - \lfloor y \rfloor \end{bmatrix},$$

where $\lfloor x \rfloor$ is the largest integer that is smaller than or equal to x, the bilinear interpolated intensity $f_{k-1}(\mathbf{r})$ is specified by

$$f_{k-1}(\mathbf{r}) = \begin{bmatrix} 1 - \theta_x \\ \theta_x \end{bmatrix}^T \begin{bmatrix} f_{00} & f_{10} \\ f_{01} & f_{11} \end{bmatrix} \begin{bmatrix} 1 - \theta_y \\ \theta_y \end{bmatrix},$$

with. The equation evaluates the 2-nd order spatial derivatives of $f_{k-1}(\mathbf{r})$ at r by means of backward differences:

$$\begin{bmatrix} g_x \\ g_y \end{bmatrix} = \begin{bmatrix} (1 - \theta_y)(f_{01} - f_{00}) + (f_{11} - f_{10})\theta_y \\ (1 - \theta_x)(f_{10} - f_{00}) + (f_{11} - f_{01})\theta_x \end{bmatrix}.$$

5.2 On the use of PCA in regression

The main idea behind the two proposed PCR procedures is the PCA of the G matrix [Jackson (1991), Jolliffe (2002)].

Each successive component explains portions of the variance in the total sample. PCA relates to the second statistical moment of G, which is proportional to $G^T G$ and it partitions G into matrices T and P (sometimes called scores and loadings, respectively), such that:

$$G = TP^T.$$

T contains the eigenvectors of $G^T G$ ordered by their eigenvalues with the largest first and in descending order. When dimensionality reduction is needed, the number of components can be chosen via examination of the eigenvalues or, for instance, considering the residual error from cross-validation [Estrela & Galatsanos (2000), Jolliffe (2002]. The PCR motion estimation algorithms will keep the PCs and use them to group DVs inside a neighborhood. The resulting clusters will give an idea about the mixture of MVs inside a mask. The formal solution PCR$_1$ may be written as

$$\hat{\mathbf{u}}_{PCR1} = P(T^T T)^{-1} T^T z, \text{ and } \hat{\mathbf{u}}_{RLS}(\Lambda) = (G^T G + \Lambda)^{-1} G^T z,$$

where a regularization matrix Λ tries to compensate for deviations from the smoothness constraint. In PCR$_1$, the scores vectors (columns in T) of different components are orthogonal. PCR$_1$ uses a truncated inverse where only the scores corresponding to large eigenvalues are included. The criteria for deciding when the PCR$_1$ estimator is superior to OLS estimators depend on the values of the true regression coefficients in the model. The previous solution can also be regularized:

$$\hat{\mathbf{u}}_{PCR2} = P(T^T T + \Xi)^{-1} T^T z.$$

with Ξ standing for a regularization matrix in the PC domain. Grouping objects can be posed as a mathematical problem consisting of finding region boundaries. Sometimes the problem is such that a sample may belong to more than one class at the same time, or not belong to any class. In this method, each class is modeled by a multivariate normal in the score space from PCA. Two measures are used to determine whether a sample belongs to a specific class or not: the leverage — the Mahalanobis distance to the center of the class, the class boundary being computable as an ellipse and the norm of the residual, which must be lower than a critical value. Figure (6) shows a set of observations plotted with respect to the first two principal components (PCs). It is likely that the four clusters correspond to four different types of DVs (see ellipses). For a big neighborhood, it could happen that these vectors would not be readily distinguished using only one variable at a time, but the plot with respect to the two PCs clearly distinguishes the populations. PCR estimates are biased, but may be more accurate than OLS estimates in terms of mean square error. Nevertheless, when severe multicollinearity is suspected, it is recommended that at least one set of estimates in addition to the OLS estimates be computed since these estimates may help interpreting the data in a different way.

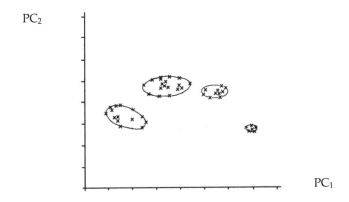

Fig. 6. An example of cluster analysis obtained by means of principal components.

When PCA reveals the instability of a particular data set, one should first consider using least squares regression on a reduced set of variables. If least squares regression is still unsatisfactory, only then should principal components be used. Besides exploring the most obvious approach, it reduces the computer load. Outliers and other observations should not be automatically removed, because they are not necessarily bad observations. As a matter of fact, they can signal some change in the scene context and if they make sense according to the above-mentioned criteria, they may be the most informative points in the data. For example, they may indicate that the data did not come from a normal population or that the model is not linear.

When cluster analysis is used for video scene dissection, the aim of a two-dimensional plot with respect to the first two PCs will almost always be to verify that a given dissection 'looks' reasonable. Hence, the diagnosis of areas containing motion discontinuities can be significantly improved. If additional knowledge on the existence of borders is used, then one's ability to predict the correct motion will increase.

PCs can be used for clustering, given the links between regression and discrimination. The fact that separation among populations may be in the directions of the last few PCs does not mean that PCs should not be used at all. In regression, their uncorrelatedness implies that each PC can be assessed independently. To classify a new observation, the least distance cluster is picked up. If a datum is not close to any of the existing groups, it may be an outlier or come from a new group about which there is currently no information. Conversely, if the classes are not well separated, some future observations may have small distances from more than one class. In such cases, it may again be undesirable to decide on a single possible class; instead, two or more groups may be listed as possible *loci* for the observation.

The average improvement in motion compensation for a sequence of K frames it turns out to be [Estrela & Galatsanos (2000)]:

$$\overline{IMC}(dB) = 10\log_{10}\left\{ \frac{\sum\limits_{k=2}^{K}\sum\limits_{\mathbf{r}\in S}\left[I_k(\mathbf{r}) - I_{k-1}(\mathbf{r})\right]^2}{\sum\limits_{k=2}^{K}\sum\limits_{\mathbf{r}\in S}\left[I_k(\mathbf{r}) - I_{k-1}(\mathbf{r} - \mathbf{d}(\mathbf{r}))\right]^2} \right\}.$$

When it comes to motion estimation, one seeks algorithms that have high values of $\overline{IMC}(dB)$. A perfect registration of motion leads to $\overline{IMC}(dB) = \infty$. Figure (7) illustrates the evolution of $\overline{IMC}_k(dB)$ as a function of the frame number for two noiseless sequences: "Foreman" and "Mother and Daughter". PCR$_2$ works outperforms the other estimators due to the use of regularization in the PC domain. Figure (8) shows the DVFs for the "Rubik Cube" sequence with SNR=20 dB.

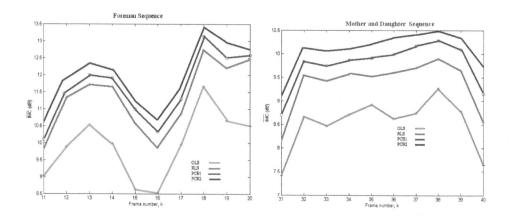

Fig. 7. Improvement in motion compensation curves for the "Foreman" and "Mother and Daughter" sequences.

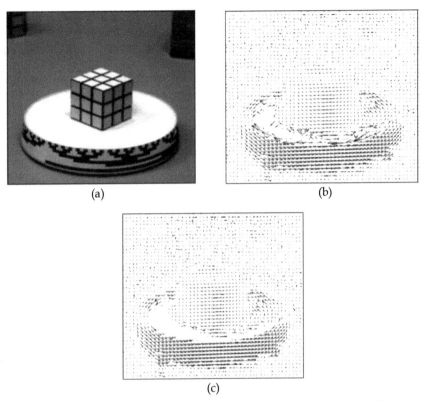

Fig. 8. Displacement field for the Rubik Cube sequence: (a) Frame of Rubik Cube Sequence; Corresponding displacement vector field for a 31×31 mask obtained by means of PCR_1 with SNR=20 dB; and (c) PCR_2, 31×31 mask with SNR=20 dB.

6. Final comments

The methods developed in this chapter allow simple and efficient computation of a few eigenvectors and eigenvalues when working with many data points in high dimensions. They rely on PPCA and the MCEM algorithm, which permit this calculation even in the presence of missing data.

The EM algorithms for PCA and KPCA derived above using probabilistic arguments are closely related to two well know sets of algorithms. The first are power iteration methods for solving matrix eigenvalue problems. Roughly speaking these methods iteratively update their eigenvector estimates through repeated multiplication by the matrix to be diagonalized. In the case of PCA explicitly forming the sample covariance and multiplying by it to perform such power iterations would be disastrous. However, since the sample covariance is in fact a sum of outer products of individual vectors we can multiply by it efficiently without ever computing it. In fact, the EM algorithm is exactly equivalent to performing power iterations for finding C using this trick. Iterative methods for partial least squares (e.g. the NIPALS algorithm) are doing the same trick for regression. Taking the singular value decomposition (SVD) of the data matrix directly is a related way to find the PS. If the Lanczos' and Arnoldi's methods are used to compute this SVD, then the resulting iterations are similar to those of the EM algorithm. The second class of methods comprises competitive learning methods for finding the PS, such as Sangers (1989) and Oja (1989) suggest. These methods enjoy the same storage and time complexities as the EM algorithm however, their update steps reduce but do not minimize the cost and so they typically need more iterations and require a learning rate parameter to be set by hand.

In this chapter, two PCR frameworks for the detection of motion fields are discussed. Both algorithms combine regression and PCA. The resulting transformed variables are uncorrelated. Unlike other works ([8, 11, 12]), the interest here is not in reducing the dimensionality of the feature space describing different types of motion inside a neighborhood surrounding a pixel. Instead, we use them in order to validate motion estimates. They can be seen as simple alternative ways of dealing with mixtures of motion displacement vectors. PCR_1 and $PCR2$ performed better than RLS estimators for noiseless and noisy images. More experiments are still needed in order to test the proposed algorithms with different types and levels of noise, so that the classification can be improved. It is also necessary to incorporate more statistical information in our models and to analyze if this knowledge will improve the outcome.

7. References

Ahn, J. H. & Oh, J. H. (2003). A constrained EM algorithm for principal component analysis, Neural Computation 15: 57-65.

Biemond, J., Looijenga, L., Boekee, D. E. & Plompen, R. H. J. M. (1987)A pel-recursive Wiener-based displacement estimation algorithm, Signal Proc., 13, pp. 399-412.

Blekas, K. Likas, A., Galatsanos, N.P. & Lagaris, I.E. (2005) A spatially-constrained mixture model for image segmentation, IEEE Trans. on Neural Networks,vol. 16, pp. 494-498.

Chattefuee, S. & Hadi, A.S. (2006) Regression analysis by example, John Wiley & Sons, Inc., Hoboken, New Jersey, USA.

Coelho, A., Estrela, V. V. & de Assis, J. (2009). Error concealment by means of clustered blockwise PCA. IEEE. Picture Coding Symposium. Chicago, IL, USA.

Dempster, A. P., Laird, N. M. & Rubin, D.B. (1977). Maximum likelihood from incomplete data via the EM algorithm. Proc. of the Royal Statistical Society B, 39:1-38.

do Carmo, F.P. , Estrela, V.V. & de Assis, J.T. (2009). Estimating motion with principal component regression strategies, MMSP '09, IEEE Int. Workshop on Multim. Signal Proc., Rio de Janeiro, RJ, Brazil, pp. 1-6.

Drew, M.S. & Bergner, S. (2004) Analysis of spatio-chromatic decorrelation for colour image reconstruction, 12th Color Imaging Conf.: Color, Science, Systems and Applications. Soc. for Im. Sci. & Tech. (IS&T)/Society for Inf. Display (SID) joint conference.

Estrela, V. V. & Galatsanos, N. (2000). Spatially-adaptive regularized pel-recursive motion estimation based on the EM algorithm. SPIE/IEEE Proc. of the Electronic Imaging 2000 (EI00), (pp. 372-383). San Diego, CA, USA.

Estrela, Vania V., da Silva Bassani, M.H. & de Assis, J. T. (2007) A principal component regression strategy for estimating motion, in Proc. of IASTED Int'l Conf. on

Visualization, Imaging and Image Processing (VIIP2007), v.2, Mallorca, Spain, pp. 1230–1234, 2007.

Estrela, V.V. & Galatsanos, N.P. (1998) Spatially-adaptive regularized pel-recursive motion estimation based on cross-validation, Proc. of ICIP-98 (IEEE Int'l Conf. on Image Proc.), Vol. III, Chicago, IL, USA, pp. 200-203.

Everitt, B. S. (1984) An Introduction to Latent Variable Models. Chapman and Hill, London.

Franz, M.O., Chahl, J.S. & Krapp, H.G. (2004) Insect-inspired estimation of egomotion. Neural Computation 16(11), 2245-2260.

Franz, M.O. & Chahl, J.S. (2003) Linear combinations of optic flow vectors for estimating self-motion - a real-world test of a neural model, Adv. in Neural Inf. Proc. Syst., 15, pp. 1343-1350, (Eds.) Becker, S., S. Thrun and K. Obermayer, MIT Press, Cambridge, MA, USA.

Franz, M.O. & Krapp, H.G. (2000) Wide-field, motion-sensitive neurons and matched filters for optic flow fields", Biological Cybernetics, 83, pp. 185-197.

Fuentes, L.M., Velastin, S.A.: (2001) People Tracking in Surveillance Applications. 2nd IEEE International Workshop on Performance Evaluation of Tracking and Surveillance, PETS2001.

Fuentes, L.M.: (2002) Assessment of image processing techniques as a means of improving personal security in public transport. PerSec. EPSRC Internal Report.

Fukunaga, K. (1990) Introduction to statistical pattern recognition, Computer Science and Scientific Computing. Academic Press, San Diego, 2 ed.

Galatsanos, N.P. & Katsaggelos, A.K. (1992) Methods for choosing the regularization parameter and estimating the noise variance in image restoration and their relation, IEEE Trans. Image Processing, pp. 322-336.

Jackson, J.E. (1991) A user's guide to principal components, John Wiley & Sons, Inc..

Jolliffe, I. T. (2002), Principal Component Analysis. Spinger-Verlag, 2 edition.

Kienzle, W., Schölkopf, B., Wichmann, & Franz, M.O. (2007) How to find interesting locations in video: a spatiotemporal interest point detector learned from human eye movements. Proc. of the 29th Conf. on Pattern Recognition, Heidelberg, DAGM 2007, 405-414. LNCS 4713, Springer, Berlin.

Kim, K. I., Franz, M. O. & Scholkopf, B. (2005) Iterative kernel principal component analysis for image modelling, IEEE Transactions on PAMI 27: 1351-1366.

Ma, X. , Bashir, F.I. , Khokhar, A.A. & Schonfeld, D. (2009) Event Analysis Based on Multiple Interactive Motion Trajectories, CirSysVideo(19) , No. 3, pp. 397-406.

McKenna, S., Jabri, S., Duric, Z., Rosenfeld, A., Wechsler, H.: (2000) Tracking Groups of People. Computer Vision and Image Understanding 80, 42–56

Rivera, L.A., Estrela, V.V. & Carvalho, P.C.P. (2004), Oriented Bounding Boxes Using Multiresolution Contours for Fast Interference Detection of Arbitrary Geometric Objects, Proc. of The 12-th Int'l Conf. in Central Europe on Computer Graphics, Visualization and Computer Vision (WSCG 2004), pp. 219-212.

Rosipal, R. & Girolami, M. (2001). An expectation-maximization approach to nonlinear component analysis, Neural Computation 13: 505-510.

Roweis, S.T. (1998) EM algorithms for PCA and SPCA, Advances in Neural Information Processing Systems, Vol. 10. pp. 626-632, MIT press.

Shawe-Taylor, I., & Cristianini, N. (2004) Kernel Methods for Pattern Analysis, Cambridge University Press, England.

Siebel, N.T., Maybank S. (2002) Fusion of multiple tracking algorithms for robust people tracking. Proceedings of ECCV 2002.

Sirovich, L. (1987) Turbulence and the dynamics of coherent structures, Quarterly Applied Mathematics, 45 (3):561–590.

Tekalp, A.M. (1995) Digital video processing, Prentice-Hall, New Jersey.

Tipping, M. E. and C. M. Bishop: (1999), Probabilistic principal component analysis, Journal of the Royal Statistical Society B 61(3), 611–622.

Vidal, R. Ma, Y. & Piazzi, J. (2004) A new GPCA algorithm for clustering subspaces by fitting, differentiating and dividing polynomials, IEEE Conf. on Comp. Vision and Pattern Recog., vol. I, pp. 510.517.

Wang, H., Hu, Z. & Zhao, Y. (2006) Kernel principal component analysis for large scale data set. Lecture Notes in Computer Science, 4113: 745-756.

Wang, Z., Sheikh, H. & Bovik, A. (2003), Objective video quality assessment, B. Furht, & O. Marques (Eds.), The Handbook of Video Databases: Design and Applications. CRC Press.

Yacoob, Y. & Black, M.J., (1999) Parameterized Modeling and Recognition of Activities, CVIU(73), No. 2, pp. 232-247.

Zhang, J., Shao, L. & Zhang, L., (2011) Intelligent video event analysis and understanding, Berlin-Heidelberg, Germany, Springer.

Zhang, J & Katsaggelos, A.K. (1999) Image Recovery Using the EM Algorithm, Digital Signal Processing Handbook, Ed. Vijay K. Madisetti and Douglas B. Williams, Boca Raton, CRC Press LLC.

Xu, L. (1998) Bayesian Kullback Ying-Yang dependence reduction theory, Neurocomputing 22 (1-3), 81-111.

Zheng, W., Zou, C. & Zhao, L. (2005). An improved algorithm for kernel principal component analysis, Neural Processing Letters 22: 49-56.

Zhao, R. & , Grosky, W.I. , (2002) Negotiating the semantic gap: from feature maps to semantic landscapes, PR(35) , No. 3, pp. 593-600.

Zelnik-Manor, L. , Irani, M. al, (2006) Statistical Analysis of Dynamic Actions, PAMI(28) , No. 9, pp. 1530-1535.

Ghahramani, Z. & Hinton, G. (1997) The EM algorithm for mixtures of factor analyzers, Technical Report CRG-TR-96-1, Dept. of Comp. Science, University of Toronto.

Ghahramani, Z. & Jordan, M.I. (1994), Supervised learning from incomplete data via an EM approach. In Jack D. Cowan, Gerald Tesauro, and Joshua Alspector, editors, Advances in Neural Inf. Processing Systems, volume 6, pages 120 -127. Morgan Kaufmann.

Wilkinson, J.H. (1965) The Algebraic Eigenvalue Problem. Claredon Press, Oxford, England.

Wold, S. et. alli., (1983) Pattern recognition: finding and using regularities in multivariate data, food research and data analysis, eds. H. Martens and H. Russwurm, London, Applied Science Publishers, pp. 147–188.

7

Principal Component Analysis in the Development of Optical and Imaging Spectroscopic Inspections for Agricultural/Food Safety and Quality

Yongliang Liu

U.S. Department of Agriculture, Agricultural Research Service
USA

1. Introduction

The American Society for Testing and Materials has defined the principal component analysis (PCA) as "a mathematical procedure for resolving sets of data into orthogonal components whose linear combinations approximate the original data to any desired degree of accuracy. As successive components are calculated, each component accounts for the maximum possible amount of residual variance in the set of data. In spectroscopy, the data are usually spectra, and the number of components is smaller than or equal to the number of variables or the number of spectra, which is less."(ASTM, 1990) Many books are available that explain the theory and mathematical basis of PCA implementation in vibrational spectroscopy covering the visible, near-infrared (NIR), mid-infrared (IR), and Raman regions (Burns & Ciurczak, 2001; Mark & Workman, 2007; Ozaki et al., 2007; Williams & Norris, 2001). This Chapter only outlines the usefulness and effectiveness of PCA in extracting valuable information from optical and imaging spectroscopy of complex agricultural /food matrixes and subsequently the development of optical and imaging spectroscopic tools for their safety and quality assessment within the recent ten years.

Interpretation of PCA pattern correctly is very important and many types of plots are available. The most frequent and essential elements are score-score and loading-loading plots. The correlations among samples are indicated by their scores (or projections) on new principal components (PCs) or the latent variables. Similar samples tend to group together in the score-score plot, and in turn, atypical samples (i.e., outliers) could be easily detected by the simple visualization.

The relationships between the new PCs and the original variables are revealed in the loading plots, in which the degree of importance decreases with the increasing PCs. A loading-loading plot displays the contributions of variables to a model directly, that is, proportional to the square of the distance from the origin. Variables far away from the origin mean their importance in discriminating the samples, whereas those close to the origin exhibit little effects on sample separation. Variables located in the same or opposite domain reflect similar or completely different information, and those with mutual locations share the common attributes.

For the comparison, standardized scores and loadings might be used, and also scores and loadings could be simultaneously displayed in biplots. In some cases, visual observation is extremely useful for elucidating the multivariate data, especially when the numbers of samples and variables are small.

2. PCA and spectroscopic sensing

Concurrent improvements in analytical techniques and data processing enable the spectroscopic devices based sensors to be more sensitive and selective, smaller, cheaper, and more robust than their laboratory version. Both optical and imaging spectroscopy is making critical judgment in assessing the safety, security, and quality aspects of agricultural and food products.

Hyper- and multi- imaging spectroscopy, which combines the features of imaging technique and vibrational spectroscopy, has been developed as an inspection means for quality and safety assessment of a number of agricultural and food products, mostly thanks to its non-invasive nature and capacity for large spatial sampling areas. However, spectral imaging technology currently has met some degree of hindrance as automated on-line systems, because current image acquisition and data analysis speeds are too slow for on-line operations.

To design rapid optical and imaging sensing systems, several essential spectral bands (usually two or three) are first sought through a variety of strategies, such as through the analysis of spectral differences in conventional visible/NIR spectra (Liu et al., 2003a), the use of PC loadings from PCA on conventional visible/NIR spectra (Windham et al., 2003), and the use of PCA on hyperspectral imaging data (Kim et al., 2002a). The selected wavebands should not only reflect the chemical /physical information in samples, but also maintain successive discrimination and classification efficiency.

Multivariate quantitative models from principal component regression (PCR) and partial least squares (PLS) require a large number of training samples to build accurate and reliable calibrations. It takes a number of initial work collecting the samples and measuring the references of targeted constitutes by the established or standard methods. In general, quantitative models will always predict reasonable values for the calibrated constitutes, given the spectra of the unknown samples are fairly similar to the training set. In other words, the reported concentrations alone will not indicate if the samples are contaminated or defected. In some scenarios, the constituents' information is not easy to determine or the samples are hard to collect. Nevertheless, the spectrum of a sample is unique to its compositions, and samples of similar compositions should have spectra that are very similar as well. Therefore, it might be possible to highlight a difference between a "good" sample and a "bad" one by only comparing the spectra with such simple methods as visual inspection or spectral subtraction. Unfortunately and apparently, these subjective methods cannot be applied for large and complicated spectral sets. Consequently, PCA based discriminant analysis, a process of classifying the samples on the basis of their spectral characteristics and also their logical assignment, is preferred and utilized considerably.

It is conceivable that the "traditional" procedure of developing the calibration models for quantitative analysis could be replaced in many applications by the operations based on

Principal Component Analysis in the Development of Optical and Imaging Spectroscopic Inspections for
Agricultural/Food Safety and Quality

127

discriminant analysis that identifies samples to be kept and those to be discarded or degraded. As an example, this could be applicable to automatic grading of grains, where such a factor as damage incurred by weather or storage conditions could result in lowering of the commercial value. Also, it could be used in screening for quality indices in plant breeding research, such as kernel texture and water absorption, by developing classes with high, medium, and low levels of the respective parameters.

2.1 Meat safety and quality

2.1.1 Diseased poultry carcasses inspection

Since the 1957 passing of the Poultry Products Inspection Act, the U.S. Department of Agriculture (USDA) inspectors have been inspecting all chickens processed at U.S. poultry plants for the indications of diseases or defects, by visually examining the exterior, the inner surface of the body cavity, and the internal organs of every chicken carcass. Especially in 1996, the USDA Food Safety and Inspection Service (FSIS) has implemented the Hazard Analysis and Critical Control Point (HACCP) program throughout the country to ensure food safety and prevent food safety hazards in the inspection of process for poultry, egg, and meat products (USDA, 1996). One requirement includes a zero tolerance standard for chickens with the signs of infections from septicemia (caused by the presence of pathogenic microorganisms or their toxins in the bloodstream) and toxemia (the result of toxins produced from cells at a localized infection or from the growth of microorganisms), and such poultry carcasses must be removed from the processing lines.

As the productivity and foreign trading arise, along with the desire from health-conscious consumers for more poultry products, workload at processing lines grows accordingly. Compared to the human inspection speed of 30-35 chickens per minute, many processing lines can run at 140 chickens a minute. To aid poultry plants in satisfying the government food safety regulations while maintaining their competitiveness and meeting consume demand, many researchers at the USDA Agricultural Research Service (ARS) during the last decade have focused on the development of new inspection technologies such as automated computer vision inspection systems.

The first-generation transportable system, developed by Chen et al. (1996), was a visible/NIR spectrophotometer based for the use in on-line and real-time classification of poultry carcasses into normal and abnormal classes. The system measured the spectral reflectance of poultry carcasses in the visible/NIR region of 471 to 963.7 nm and an optimal neural network classifier was used for the separation of poultry carcasses into two categories at the average accuracy of 97.4%. In order to facilitate the selection of optimal spectral regions and imaging camera and also further to enhance the effective discrimination, a number of strategies were taken, including two-dimensional (2D) correlation analysis on small samples and PCA approach on large sample sets. For example, Liu et al. (2000) first applied 2D correlation analysis to characterize the spectral variations between wholesome and unwholesome (diseased) chicken meats. In this work, they proposed the spectral band assignments for deoxymyoglobin (445 nm), oxymyogobin (560 nm), and metmyoglobin (485 nm) species that are mainly responsible for the meat color. Also, they concluded that the three pigments co-exist in all fresh-cut wholesome and diseased meats, but with a clear indication that wholesome meats have more variation in deoxymyoglobin and oxymyogobin and less metmyoglobin than do diseased meats.

With the consideration of significant absorptions in the 400-500 nm range, the follow-up experiments had been conducted by the use of optical spectrometers capable of scanning the spectra in much shorter wavelength up to 400 nm (Liu & Chen, 2003). They compared the PCA/SIMCA (soft independent modeling of class analogies) 2-class (wholesome vs. unwholesome) classification results between the visible spectral region (400-700 nm) and the entire 400-2500 nm spectral region. It was observed that the differentiation in the 400-700 nm range is almost as effective as that from the entire spectral region with a correct separation of 93.8%. Their findings echoed the observation that visible spectroscopy is useful in studying the variation of meat color.

Chao et al. (2003) took a different visible/NIR spectroscopic system and discrimination strategy intended to further understand how the wholesome and unwholesome carcasses could be separated. Using PCA and a linear discriminant function, they reported the best visible/NIR classification model with correctly classified 100%, 90.0%, and 92.5% rate of the whole (skin and meat) samples for wholesome, septicemia, and cadaver categories, respectively. Examination of the PCA loadings for the whole samples suggested that the better discrimination of whole samples was dependent on spectral variation related to different forms of myoglobin present in the chicken meat, i.e. deoxymyoglobin, metmyoglobin, and oxymyoglobin. In particular, key wavelengths were identified at 540 and 585 nm, which have been identified as oxymyoglobin bands, for PCs 1 and 2; 485 nm, metmyoglobin, for PC 3; and 440 nm, deoxymyoglobin, for PC 8.

The second-generation system involved a development of hyper- and multi- spectral imaging systems. At earlier stage, Park et al. (1996) first described the spectral image characterization of poultry carcasses at a variety of conditions based on the gray-scale intensity, Fourier power spectrum, and fractal analysis, and then reported a neural network classifier performance of 91.4% accuracy for the separation of tumorous carcasses from normals based on the images scanned at both 542 and 700 nm wavelengths.

Through the accumulated knowledge from extensive studies over the years, the line-scan image camera capable of scanning the visible region of 389 to 753 nm was adopted for the use in the most recent investigations, as reported by Chao' team (Chao et al., 2007; Yang et al., 2010). They documented the achievement of 90.6% classification accuracy for wholesome birds and 93.8% accuracy for diseased birds in the calibration data set as well as 97.6% accuracy for wholesome birds and 96.0% accuracy for diseased birds in the testing data set during the in-plant trial. Finally, they concluded that the hyper- and multi- spectral line-scan imaging systems can be used for automated on-line inspection of chicken carcasses for the detection of systemically diseased birds on high-speed processing lines, ultimately to increase inspection efficiency, reduce labor and cost, and produce significant benefits for poultry processing plants.

2.1.2 Fecal contaminated poultry carcasses inspection

Contamination of meat and poultry with food-borne bacterial pathogens can potentially occur as a result of exposure of the animal carcass to fecal materials during or after slaughter. Microbial pathogens can be transmitted to humans by consumption of contaminated undercooked or mishandled meat and poultry. Bacterial pathogens in food cause an estimated 76 million cases of human illness and up to 5000 death annually in the

Principal Component Analysis in the Development of Optical and Imaging Spectroscopic Inspections for
Agricultural/Food Safety and Quality

129

U.S. To ensure a healthy and safe meat supply, the USDA FSIS has established a zero tolerance policy to minimize the likelihood of bacterial pathogens on the surfaces of meat and poultry carcasses during the slaughter. Applicable HACCP programs require individual meat processor to identify all food safety hazards in the process and to identify critical control points adequate to prevent them. Preventing carcasses with visible fecal contamination from entering the chlorinated ice water tank (chiller) is critical for preventing cross-contamination of other carcasses. Thus, the final carcass wash to remove all surface-adhering feces, before entering the chiller, has been adopted by many poultry processors as an HACCP system critical control point.

Compliance with zero tolerance in meat processing is routinely verified through human visual observation where the criteria of color, consistency, and composition are used for the identification. Trained inspectors use the established guidelines to verify that carcasses with visible fecal contaminations must be removed prior to entering the chillers. Current visual inspection is both labor intensive and prone to both human error and inspector-to-inspector variation. Therefore, investigators at the USDA ARS have been developing multi- and hyper- spectral imaging systems in real-time on-line detection of fecal contaminated chicken carcasses (Lawrence et al., 2003; Park et al., 2004).

Aiming to improve the detection performance of imaging systems with optimum settings, Windham et al. (2003) reported the use of multivariate data analysis on visible/NIR reflectance spectra to determine specific wavelengths for analyzing the imaging spectra from relative intensities of PC loadings. As a complementary approach, Liu et al. (2003a) presented a novel methodology to analyze and then classify the visible/NIR spectra of uncontaminated chicken skins and pure chicken feces as well as hyperspectral imaging spectra of fecal contaminated chicken skins. By examining the spectral difference, they identified several characteristic bands and subsequently developed simple two- or three-band subtraction and ratio algorithms. Their results revealed that both algorithms could be used to perform the classification analysis between skins and feces class with a great success, which was in good agreement with 2-class PCA/SIMCA models (skins vs. feces).

Recently, Park et al. (2011) have updated their efforts in commercializing the imaging system. Latest research demonstrated the feasibility of the system in terms of processing speed and detection accuracy for a real-time, in-line fecal detection at current processing speed (at least 140 birds per min) of commercial poultry plant. The preliminary results showed the real-time hyperspectral imaging system could detect small amount (about 10 mg) of fecal and ingesta contaminants, and the system performance could be improved by optimizing line lighting system especially NIR bands for quality images and additional spectral images to minimize false positive detection errors.

2.1.3 Sanitation efficiency at poultry processing plant

Relative to possible fecal contamination of meats during the slaughter processing, there is also a concern that meats might be contaminated from fecal remains on the surfaces of equipment, utensils, and walls at slaughter plants. To this regard, the USDA FSIS's mandatory HACCP systems require all meat and poultry plants to develop written sanitation standard operating procedures to show how they will meet daily sanitation requirements. This is important in reducing pathogens on poultry because unsanitary practices in plants increase the likelihood of product cross-contamination. Thus, slaughter plants are required not only to document daily records of completed sanitation standard operating procedures, but also to undergo hands-on sanitation verification by the USDA FSIS inspectors.

Evaluations and inspections of sanitation effectiveness are usually performed through one or more of the following methods; organoleptic (e.g., sight and feel), chemical (e.g., checking the chlorine level), and microbiological (e.g., microbial swabbing and culturing of product contact surface). As poultry feces are the most likely source of pathogenic contamination, the USDA FSIS inspectors use the established guidelines to identify fecal remains on the surfaces of equipment, utensils, and walls at slaughter plants. Certainly, the inspection duty is not easy, both labor intensive and prone to human error. Scientists and engineers at the USDA ARS have been looking into low-cost, reliable, and portable sensing devices, such as head-wear goggles and binoculars, by extending the scope of hyper- and multi- spectral reflectance and fluorescence imaging systems. One key factor in successful applications is to use a few essential spectral bands that meet the discrimination expectations.

On a direct analysis of visible and NIR spectral differences between feces/ingesta objectives and rubber belt/stainless steel backgrounds, Liu et al. (2006a) identified a number of significant bands and then developed simple three-band ratio algorithms for discriminant analysis. They observed that the three-band based algorithms could classify feces/ingesta objectives from rubber belt/stainless steel backgrounds with a success of over 97%, which was at least the same accuracy as those from the 2-class SIMCA models (feces/ingesta objectives vs. rubber belt/stainless steel backgrounds). Meanwhile, PCA was performed on both spectral sets, and the score-score plot showed a clear separation between feces/ingesta objectives and rubber belt/stainless steel backgrounds. However, the optimal loadings did not provide any specific characteristic bands that could further improve the classification rate. The finding of three visible or NIR bands is most promising in the development of simple goggle and binocular sensing system for in-situ inspection of fecal and ingesta contaminants at slaughter plants.

2.1.4 "Tender" / "Tough" poultry meat and meat quality

To provide poultry processors with accurate, reliable, and rapid information on the evaluation of meat quality attributes and, further, to facilitate the efficiency of new processing techniques, food scientists have been focusing on the relationships between sensory attributes and changes in the production process (Lyon & Lyon, 1991, 1996). In these pioneering studies, trained sensory panels and instrumental measurements were used together to draw conclusions and make decisions about meat quality. Instrumental methods, such as the Warner-Bratzler (W-B) shear force, can measure characteristics that are directly related to the physical components of meat products and can provide reliable information about meat quality. However, human subjects go beyond the physical components to describe a wide range of factors involved in mastication and afterfeel/aftertaste sensations, such as appearance, flavor, and texture. Sensory panels provide complementary information to instrumental method, and neither can be replaced. For example, instruments do not account for the juiciness and other moisture-related characteristics that panelists may perceive while chewing, and panels may identify and quantify more specific texture attributes that are not measured instrumentally. Meanwhile, relationships might exist between instrumental measurements and sensory panel evaluations. Previous studies have established a range of instrumental shear force values corresponding to different portions of the consumer texture scale, which enables commercial processors to relate the meaning of instrumental shear values to terms of relative toughness/tenderness of broiler breast meats (Lyon & Lyon, 1991).

Although instrumental W-B shear force measurement and sensory evaluation techniques can provide reliable information about poultry meat quality, it is destructive, time-consuming, and unsuitable for on-line application. The development of fast, non-destructive, accurate, and on-line / at-line techniques is critical to increase processing efficiency. Visible / NIR spectroscopy could form the basis for such techniques due to the speed, ease of use and less interference from color of meat samples.

The preliminary study, reported by Liu et al. (2004b), suggested that visible/NIR technique might have the potential to predict W-B shear force value, color, pH, and sensory characteristics in broiler muscles. As expected, the predictive models of meat color indices (L*, a*, b*), pH, and W-B shear force have better accuracies than those of individual sensory attributes. From visible/NIR predicted tenderness values in PLS model, breast samples were classified into "tender" and "tough" classes with a correct classification of 74.0% if the boundary was set to be 7.5 kg. As an alternative, a model based on PCA/SIMCA of measured shear force values as an indication of tenderness was attempted, and it showed nearly the same classification success.

A variety of chemical, physical, color, and sensory analyses are necessary to completely describe the characteristics of meats. Each type of analysis contributes specific and important information on overall meat quality. In other words, a meat sample might be characterized with more than one technique, resulting in many diverse parameters (variables). Generally, it is difficult to obtain a comprehensive overview of many meat samples with a number of variables. Hence, it might be useful to reduce the number of variables to describe the meats. As a strategy, Liu et al. (2004a) applied PCA to characterize the variations of a total 24 variables representing the objective and sensory properties of broiler breast meats deboned at different times. They observed several significant correlations among these variables, and W-B shear force had high positive correlations with 5 sensory texture attributes. Although PCA score plot showed no clear separation of the breast muscles deboned at different postmortem times, it could be still possible to differentiate them. The loading biplot suggested that 18 variables were effective in meat differentiation, including W-B shear force. However, the means to obtain either objective or sensory properties are destructive, time-consuming, and unsuitable for meat quality grading at large-scale operation or on-line implementations. As a part of conclusion, they suggested the development of fast, nondestructive, and on-line/at-line optical or imaging techniques for qualitative and quantitative determination of poultry meat eating qualities.

2.2 Grain safety and quality

2.2.1 DON contaminant screening

Deoxynivalenol (DON), also known as vomitoxin, is a type B trichothecene mycotoxin. It is one of major secondary metabolites produced by fungi of the *Fusarium* genus and occurs predominantly in grains such as wheat, barley, and corn (Leonard & Bushnell, 2004). The presence of DON has been reported to cause the quality degradation of grain and also a variety of very real toxic effects in humans and livestock who have consumed DON contaminated grain products. Authorities in a number of countries and organizations have established regulatory levels or guidelines for DON in food and feed. For example, the U.S. Food and Drug Administration (FDA) has proposed advisory levels for DON at 1 mg kg^{-1} for finished wheat products for humans, 5 mg kg^{-1} for swine and other non-ruminants, and 10 mg kg^{-1} for cattle and poultry feed. Though the milling process typically reduces DON concentration by approximately one-half, subsequent baking or heating processes cannot destroy DON toxin due to its thermal stability.

A number of analytical methods, such as thin-layer chromatography (TLC), gas chromatography (GC), high-performance liquid chromatography (HPLC), mass spectrometry (MS), and GC- / HPLC- coupled MS, have been developed to measure DON concentration in grain. Clearly, these traditional methods involve expensive and time-consuming steps, including solid-phase extraction, separation, detection, and sample cleanup. New development of biotechnological approaches (e.g., biosensors and immunoassays) has been reported for rapid and specific detection of DON at trace levels, but these attempts still involve extraction and washing steps as well as the extra time for binding process.

Fast DON screening requires minimal sample preparation (e.g., to avoid the extraction / centrifugation), permits routine analysis of a number of samples with minimal use of reagents, requires fewer procedures, and is easy to operate. Vibrational spectroscopy is an alternative approach, since it can be applied directly to the solid grain in the state of single-kernel and ground without any DON extraction steps (Delwiche & Hareland, 2004; Delwiche, 2008). In general, grain contains a large portion of moisture that, in turn, yields intense and broad water bands in both the IR and NIR regions and can substantially hide other useful bands attributable to protein and carbohydrate species. Conversely, the Raman technique, which is based on the polarizability of bonds and not their dipoles like IR, is insensitive to water and provides fewer overlapping bands.

Due to DON toxin at the ppm concentration level, it is unlikely that the Raman method is directly sensitive to differences in DON levels. Rather, as DON is produced as a metabolite of *Fusarium* fungi during the growth of grain, it causes side effects on chemical, physical, color, and structure of grain. In turn, these effects could result in minor but significant changes in relative intensity, position and shape of Raman bands between low DON grain and high DON grain. Based on such spectral distinctions, Liu et al. (2009) suggested the use of two Raman bands near 1560 and 904 cm^{-1} in creating simple intensity-intensity plot for discrimination analysis. Their observation from a limited set of samples revealed that the simple intensity-intensity algorithm could be used to classify low DON grains from high DON ones, which were well confirmed by the PCA/SIMCA models. Notably, the use of Fourier transform (FT) methodology and a 1064 nm NIR excitation laser provides precise wavenumber measurement and good-quality Raman spectra by reducing the interference from fluorescence and photodecomposition of chemical components in wheat and barley.

2.2.2 *Fusarium* damage assessment

Fusarium head blight (scab) is a worldwide fungal disease that affects the small grains such as wheat and barley. Affecting the spikelets during plant development, the fungus causes a reduction of yield and further compromise the grain quality. Importantly, secondary metabolites that often accompany the fungus, such as DON, are health concerns to humans and livestock. Conventional grain inspection procedures for *Fusarium* damage are heavily reliant on human visual analysis. As an inspection alternative, Delwiche et al. (2011a) investigated the potential of hyperspectral image systems (1000 -1700 nm NIR vs. 400-1000 nm visible) in the detection of *Fusarium*-damaged wheat kernels. On a limited set of wheat samples that their conditions were subjectively assessed and also using a linear discriminant analysis (LDA) classifier, they found that hyperspectral imaging in either visible or NIR regions was able to discriminate *Fusarium*-damaged kernels from sound kernels at an average accuracy of approximately 95%.

2.2.3 Grain discrimination

Canada is one of the most wheat producers and exporters in the world. In Canada, wheat is classified based on color (red or white), hardness (soft or hard), and growing season (winter or spring). A specific wheat class is used as a primary raw material for specific products, such as Canada Western Red Spring (CWRS) wheat is processed for loaf bread. At present, visual method which requires extensive training and experience, is commonly used to identify wheat classes in grain handling facilities. A machine vision technique has been used to differentiate two wheat classes (Canada Western Red Spring vs. Canada Western Amber Durum). However, human inspection cannot be utilized for identifying wheat of different moisture levels because of subjectivity of the method. Higher moisture wheat (>15%) needs to be dried to an optimal level (12-13%) to be stored safely so as to prevent its spoilage and/or sprouting prior to processing.

A NIR hyperspectral imaging system (960–1700 nm) has been explored to identify five western Canadian wheat classes at varying moisture levels (Mahesh et al., 2011). Besides the generation of scores images and loadings plots from PCA, the linear and quadratic discriminant analyses were used to classify wheat classes giving accuracies of 61–97 and 82–99%, respectively, independent of moisture contents. They also observed that the linear discriminant analysis (LDA) and quadratic discriminant analysis (QDA) could classify moisture contents with classification accuracies of 89–91 and 91–99%, respectively, independent of wheat classes. Once wheat classes were identified, classification accuracies of 90–100 and 72–99% were observed using LDA and QDA, respectively, when identifying specific moisture levels. From this study, it was concluded that hyperspectral imaging technique can be used for rapidly identifying the wheat classes even at varying moisture levels.

2.2.4 Differentiation of waxy wheat

Wheat (*Triticum aestivum L.*) breeding programs are currently developing varieties that are free of amylose (waxy wheat), as well as genetically intermediate (partial waxy) types. Successful introduction of waxy wheat varieties into commerce is predicated on a rapid methodology at the commodity point of sale that can test for the waxy condition. In meeting this trend, Delwiche et al. (2011b) examined the ability of NIR reflectance spectroscopy to differentiate the starch waxy genotypic groups in hard winter wheat breeders' lines representing all eight genotypic combinations. By applying the LDA of PC scores, they noted that fully waxy wheat is identifiable at typically 90-100% accuracy. Since the fully waxy trait can be easily identified by NIR, regardless of the genetic background (population) within which it resides, hopefully breeders could easily use NIR to select waxy lines from early generation materials. Further, they pointed out the potential that end users could also utilize NIR to differentiate waxy crops at harvesting sites.

2.3 Fruit and vegetable safety

2.3.1 Fecal contaminant inspection on apples

Apple products could be contaminated with bacteria pathogens due to the contact with fecal materials in the phases of growing and harvesting. Animal feces are the most likely source of pathogenic *E. coli* O157: H7 contamination. In addition, the potential of contamination increases with physical damages on apples, such as lesions and bruises, which provide a site for bacterial growth. Cleaning processes can reduce, but are unlikely to eliminate, pathogens from the surfaces of produce even if antimicrobial chemicals are contained in the wash water. Bacterial pathogens can be presented in contaminated apples or raw (unpasteurized) apple juice / cider. There have been several reported foodborne illness outbreaks attributed to unpasteurized apple juice and cider. These outbreaks have raised the concerns of public health officials and apple cider / juice producers.

Responding to this interest, the U.S. FDA has issued an HACCP system to minimize the likelihood of any pathogens in fruit juices and identified an urgent need to develop methods for detecting fecal matters on apples. One element of the guidelines, on good agricultural practices (GAPs) and good manufacturing practices (GMPs) for fruits and vegetables, suggests the removal of fecal contaminated apples before entering the washer tank.

At present, inspection of fecal contamination is through visual observation over an inspection table. Inspectors use the GMPs guidelines to prevent apples with visible fecal contaminants from entering the next step. Current visual inspection is labor intensive and prone to human error and inspector-to-inspector variation. A research team led by Kim has been developing both hyperspectral reflectance and fluorescence image systems for the detection of fecal contaminated apples (Kim et al., 2002a, 2002b; Yang et al., 2012). In the systematic approach, they utilized a hyperspectral reflectance imaging technique in conjunction with the use of PCA to define several optimal wavelength bands. The investigation illustrated that, with the use of the PCA, high spectral dimension reflectance image data were reduced to several optimal wavelengths (multispectral) images. Also, they suggested three visible–NIR bands that could potentially be implemented in multispectral imaging systems for detection of fecal contamination on apples.

In supporting the selection of minimum and effective spectral bands in the image processing, Liu et al. (2007) characterized the distinctions in Region of Interest (ROI) spectral features between fecal contaminated areas and uncontaminated apple surfaces, and found the occurrence of large spectral differences in the 675-950 nm visible/NIR region, which provided the basis for developing universal algorithms in the detection of fecal spots. Comparison of a number of processed images (including those from PCA), they determined that a dual-band ratio ($Q_{725/811}$) algorithm could be used to identify fecal contaminated skins effectively. The observation was most important as the two bands are away from the absorptions of natural pigments (such as chlorophylls and carotenoids), and hence can reduce the influence from color variations due to different apple cultivars.

Complementary to acquire images and also to overcome the low sensitivity of detecting thin fecal smears (or low concentrations) in visible/NIR region, Kim et al. (2002b) proposed the use of hyper- and multi- spectral fluorescence imaging for classification of fecal contaminated apples. They utilized both PCA and visual examination to determine the optimal bands that allow the effective recognition of fecal contaminations on apple surfaces.

2.3.2 Fecal contaminant detection on cantaloupes

Since its first notice in early September 2011, the *listeria* outbreak linked to a crop of cantaloupes has now claimed the lives of at least 23 people and sickened more than 109 people across 23 U.S. states as of mid-October. The deadliest foodborne illness in more than a decade matches the death toll from a multi-state *listeria* outbreak linked to hot dogs and deli turkey that started in 1998 and stretched into 1999.

Cantaloupes become contaminated with pathogens through direct contact with manure, contaminated soil, animals or humans during any stage of the food-handling chain, including the growing and harvesting operations as well as while in the processing plants. In general, the pathogens originate from the intestinal tracts of animals and humans, thus making fecal matter a major source of contamination. For instance, contaminated cantaloupes were found to be responsible for 2 deaths and 18 hospitalizations due to *Salmonella* bacteria between 2000 and 2002, and a thorough investigation revealed the cause of unsanitary conditions in processing and packaging plants (Anderson et al., 2002).

In 2005, Kim's research group probed the feasibility of hyperspectral fluorescence images in detecting fecal spots on cantaloupes that were artificially contaminated with bovine feces at varying concentrations (Vargas et al., 2005). To improve the detection algorithms, they presented several image processing tactics (such as single-band and two-band ratio images) and found the potential of the PCA processing of hyperspectral images in the detection of fecal contaminated spots (a minimum of 16-μg/mL dry fecal matter) on cantaloupes with minimal false positives. With the examination of PC weighing coefficients, they identified several dominant wavelengths that could be implemented to a multispectral imaging system for further on-line applications.

2.3.3 Bruise detection

Bruises are of great concern to the fruit and vegetable industry and the retailer because they lower the quality grade of the produce and can cause significant economic losses. Bruising normally happens to the tissue beneath the fruit skin. After the fruit tissue is damaged, its cells are initially filled with water and turn brownish. As time elapses, the damaged cells start to lose moisture and eventually become desiccated. It is a challenging task to detect bruises on the fruit, in part because of the presence of fruit skin and in part because detection accuracies are affected by factors such as time, bruise type and severity, apple variety, and fruit pre– and post–harvest conditions. Not long ago, Lu (2003) developed a NIR hyperspectral imaging based bruise detection system for detecting bruises on apples in the spectral region between 900 nm and 1700 nm. His results indicated that the spectral region between 1000 nm and 1340 nm was most appropriate for bruise detection. He also observed that bruise features changed over time from lower reflectance to higher reflectance, and the rate of the change varied with fruit and variety. Using both PC and minimum noise fraction transforms, his system was able to detect both new and old bruises, with a correct detection rate from 62% to 88% for Red Delicious and from 59% to 94% for Golden Delicious. The optimal spectral resolution for bruise detection was between 8.6 nm and 17.3 nm, with the corresponding number of spectral bands between 40 and 20.

Later, Ariana et al. (2006) applied the same imaging system to capture hyperspectral images from pickling cucumbers at 0–3, and 6 days after they were subjected to dropping or rolling under load which simulated damage caused by mechanical harvesting and handling systems. PCA, band ratio, and band difference were applied in the image processing to segregate bruised cucumbers from normal cucumbers. They reported that bruised tissue had consistently lower reflectance than normal tissue and the former increased over time. Best detection accuracies from the PCA were achieved when a bandwidth of 8.8 nm and the spectral region of 950–1350 nm were selected. The detection accuracies from the PCA decreased from 95 to 75% over the period of 6 days after bruising, which was attributed to the self-healing of the bruised tissue after mechanical injury. The best band ratio of 988 and 1085 nm had detection accuracies between 93 and 82%, whereas the best band difference of 1346 and 1425 nm had accuracies between 89 and 84%. From general classification performance analysis, they concluded that the band ratio and difference methods had similar performance, and both were better than the PCA.

2.3.4 Chilling injury inspection

It is well-known that many fruits and vegetables are sensitive to chilling and are damaged by low temperatures during the storage and transportation process. Cucumber is one such produce, apt to suffer chilling injury from relatively short periods of time at low temperatures. Extensive decay occurs when chilling injured cucumbers are returned to warmer temperatures, and damaged areas can become locations for further fungal decay and bacterial infection. Accumulated bacterial pathogens from these areas can be harmful to humans by consumption of uncooked or mishandled cucumbers.

Chilling injury has been of great attention to the fruit and vegetable industry, and a diversity of methods has been developed to reduce the occurrence of chilling injury for cold-sensitive produces (Wang, 1993). These techniques include low temperature preconditioning, intermittent warming, waxing, genetic modification, and chemical treatments. Nevertheless, the development of rapid, non-destructive, and accurate methodologies that are suitable for on-line and at-line operations is critical to increase the efficiency of cucumber safety/quality evaluation. Hyperspectral imaging spectroscopy can be the basis for the development of such techniques, thanks to its non-invasive nature and capacity for large spatial sampling areas.

Usually, the amount of information contained in hyperspectral images is excessive and redundant, and data mining for waveband selection is needed. In the applications such as fruit and vegetable defect inspections, effective spectral combination and data fusing methods are required in order to select a few optimal wavelengths without losing the crucial information in the original hyperspectral data. Cheng et al. (2004) proposed a novel method that combines PCA and Fisher's linear discriminant (FLD) method to show that the hybrid PCA–FLD method maximizes the representation and classification effects on the extracted new feature bands. The method was then applied to the detection of chilling injury on

Principal Component Analysis in the Development of Optical and Imaging Spectroscopic Inspections for
Agricultural/Food Safety and Quality

137

cucumbers. Based on tests on different types of samples, their results showed that this new integrated PCA–FLD method outperforms the PCA and FLD methods when they were used separately for classifications.

As a differing approach, Liu et al. (2006b) tested a variety of image processing and visually compared the detection efficiency of chilling injury. Firstly, they examined the ROI spectral features of chilling injured areas that showed a reduction in reflectance intensity during multi-day post-chilling periods of room temperature (RT) storage. Next, they determined the large spectral difference between good-smooth skins and chilling injured skins occurred in the 700 to 850 nm visible/NIR region. Then, a number of data processing methods, including simple spectral band algorithms and PCA, were attempted to discriminate the ROI spectra of good cucumber skins from those of chilling injured skins. The observation indicated that using either a dual-band ratio algorithm ($Q_{811/756}$) or a PCA/SIMCA model from a narrow spectral region of 733 to 848 nm could detect chilling injured skins with a success rate of over 90%. Further, they applied the dual-band algorithm to the analysis of images of cucumbers at different conditions, and the resultant images showed more correct identification of chilling injured spots than PCA method. The results also suggested that chilling injury was relatively difficult to detect at the stage of the first 0 to 2 days of post-chilling RT storage, due to insignificant manifestation of chilling induced symptoms.

2.4 Single *Bacillus* spores detection

Under stressed conditions such as lack of key nutrients, certain *Bacillus* cells in the vegetative state will spontaneously develop into a dormant state known as an endospore. The spore is organized into a series of concentrically arranged structures, each of which contribute in a different way to resist against environmental stresses such as heat, radiation, desiccation and chemical disinfectants. Detection of *Bacillus* spores is of considerable importance in agricultural and food industries, since the ubiquity of these spore-forming bacteria allows it to potentially threaten the safety of a wide range of foods, including dairy products, meats, cereals, vegetables, spices, and ready-to eat meals. In addition, Bacillus spores can survive standard processing and sanitation treatments for foods and food processing equipment.

There has been growing interest in the applications of optical methods such as Raman spectroscopy for microbial characterization. Major advantages of Raman approach are that samples can be analyzed with minimum preparation in aqueous state and measured non-destructively on-line, and in real time. However, Raman signals are normally very weak and can only be used for bulk samples or concentrated solutions. The problems can be overcome by using a much more sensitive method, namely surface enhanced Raman spectroscopy (SERS), which could reach the limit of detection (LOD) to a single spore or cell. When the targeted molecules are attached to noble metal substrates (typically, Au or Ag), SERS effect will occur with the enormously enhanced Raman signals.

With the use of gold SERS-active substrates, He et al. (2008) were able to observe distinct spectral differences among five different *Bacillus* spores at single spore level. In the following, hierarchical cluster analysis (HCA) and PCA were applied and the results showed clear data segregations at the species level between five Bacillus spores. The corresponding PC values indicated that the Raman range between 900 and 1200 cm^{-1}

contributed significantly to the total data variance in the PCA plot. In particular, a prominent band of dipicolinic acid (DPA) was observed at 998 cm^{-1} and served as a biomarker for bacterial spores. Their study demonstrated that SERS method is a promising tool for rapid, ultra-sensitive, and selective detection of bacterial spores in foods and other complex biological matrices.

2.5 Cotton quality grading

Micronaire property has been recognized as one of key cotton quality indices for fiber classers and processors. It is a measure of fiber fineness and maturity, and is determined by measuring the air permeability of a constant mass of cotton fiber compressed to a fixed volume. Previous studies have implied the ability of NIR technique to determine cotton micronaire with a relatively high degree of success (Liu et al., 2010). Apparently, NIR predicted micronaire values could be near the boundaries separating three cotton classes of "Discount Range", "Base Range", and "Premium Range", which might be a problem and the source of error during the cotton classification.

Following the assignment of fibers into "Discount Range", "Base Range", and "Premium Range" classes according to predicted micronaire values from optimal PLS model, 16, 12, and 9 samples were correctly classified at respective classes of 18, 13, and 10 cotton samples, with a 90.4% of overall classification (Liu et al., 2010). For a comparison, the 3-class based SIMCA/PCA discriminant models were created, and a better separation power than respective PLS model was observed.

3. PCA and 2D correlation analysis

Two-dimensional (2D) correlation spectroscopy, a universal and modern technique of vibrational spectral analysis, was originally developed as 2D IR correlation spectroscopy by Noda (1986). In this initial concept, a system was excited by an external perturbation that includes dynamic fluctuations of IR signals, and then a simple cross-correlation analysis was applied to sinusoidally varying dynamic IR response to generate a set of 2D IR correlation spectra. Years later, Noda (1993) introduced a more applicable and simple mathematical formalism to perform the generalized 2D correlation analysis, which has been considerably and successfully applied not only to a variety of optical spectroscopic techniques (IR, NIR, Raman, visible, fluorescence), but also for a number of different types of simple external perturbations (electrical, thermal, magnetic, chemical, acoustic, mechanical, spatial positions etc) and waveforms. When an external perturbation was applied to a sample system, specific components within the system could be selectively excited and, subsequently, be monitored with many different types of electromagnetic probes. Over the period, generalized 2D correlation spectroscopy has been established as a viable means to analyze and extract useful information from conventional one-dimensional (1D) spectral data.

Major advantages of generalized 2D correlation spectroscopy include the enhancement of spectral resolution by spreading peaks over the second dimension, the band assignments through the correlation analysis, and probing the complex sequence of events arising from the changes in a system. To obtain generalized 2D correlation spectra and also to interpret them in a reasonable manner, a limited number of spectral data was arranged in an increasing or decreasing variable. On the other hand, one of the specific challenges might be

Principal Component Analysis in the Development of Optical and Imaging Spectroscopic Inspections for
Agricultural/Food Safety and Quality

139

how to implement generalized 2D correlation analysis in large and diverse spectral sets, which are common in chemometric model developments and actually include multivariate variations in chemical and physical components, and to acquire useful information from them. As an approach to large samples with at least one known attribute, the use of the average spectra that had close physical or chemical values has been reported (Liu et al., 2004c). Being another method to explore the variations within a diverse spectral set, PCA was utilized to classify the spectra of samples (Liu & Chen, 2000; Kokot et al., 2002). Further, 2D characterization was attempted to understand the PC loading spectra on large spectral data set with multi-variables (Liu et al., 2003b).

3.1 Cluster of samples

Most 2D applications are limited to one perturbation (or variable) dependent spectral intensity change within a simple system. To extend the scope of 2D analysis to more complicated biological samples, Liu & Chen (2000) proposed a strategy of utilizing PCA procedure to identify two NIR spectral clusters of chicken meats at different periods during cold storage for the first time in 2000, in which one group representing the storage process from day 2 to day 9 and another one from day 10 to day 18. It opens a way to implement 2D study in more applications and, anticipated, this PCA process has been successfully attempted to other agricultural products, such as cotton fibers at various elevated temperatures (Kokot et al., 2002).

In the most recent 2D study of characterizing the attenuated total reflection (ATR) spectral intensity fluctuations of immature and mature cotton fibers, their conditions were identified subjectively and then were verified by the PCA (Liu et al., 2011a). As a traditional practice, the cotton bolls were taken at various days post-anthesis (DPAs) to unravel a number of interests in structure, maturity, and physical properties among developing fibers. Abidi et al. (2010) have employed PCA to analyze the FT-IR/ATR spectra of fibers as a function of developmental DPAs, and observed that the PC1 scores, in general, increased with DPAs for two cotton varieties. Although the PC1 scores were not linear with DPAs, they identified two groups of spectra with negative PC1 scores for shorter DPAs and positive PC1 scores for longer DPAs, and further concluded the transition phase at 17 and 21 DPAs for the respective cotton varieties. The PCA results were consistent between two sampling approaches, visual inspection (Liu et al., 2011a) and differing DPAs (Abidi et al., 2010).

The unique bands identified through the 2D study could be used to interpret two key bands at 1032 and 956 cm^{-1} that were utilized to develop simple algorithm for the classification of immature fibers from mature ones (Liu et al., 2011b). For example, by setting a ratio value of 0.4, it is possible to detect immature fibers positively with over 95% success rate, and, furthermore, the degree of fiber maturity was assessed.

3.2 PC loadings

Liu et al. have systematically investigated the 2D visible/NIR spectral correlation analysis of chicken breasts under various conditions, such as cooking, thawing, and cold storage as well as diseases (Liu & Chen, 2000, 2003; Liu et al., 2000). These results have shown that the visible bands identified through the 2D studies have been found to be useful as an indicator of meat color variation during cooking, irradiation and cold storage. They have displayed the significance of 2D approach in analyzing overlapped and broad bands of meat products, and also established the relationship between spectral absorption and meat color structure that has been indicative in creating the next-generation sensing devices.

To understand the variations within the large spectral set, Liu et al. (2003b) analyzed the loadings spectra of PCA. They presented three examples of visible/NIR spectra of chicken breast muscles under a variety of treatments, as muscles are one of the complicated agricultural commodities that vary greatly in color, chemical, physical and sensory attributes from one portion to another. The 2D results indicated that characteristic bands from the loadings spectra are in good agreement with those from a small number of spectra induced by simple external perturbations in previous investigations. They concluded that although some advantages of 2D correlation analysis (such as sequential changes in intensity) were not available, it might still be useful for the interpretation of large and complex spectral data set with multi-variable variations.

4. Conclusions

The Chapter reviews the recent developments of PCA in optical and imaging spectroscopy for agricultural and food safety and quality. Food safety is one of most important issues for public health, and authorities have zero tolerance performance standards for various food products. Driven by this increasing interest of protecting food supply, the ability of optical and imaging spectroscopic techniques to rapidly, routinely, potentially to be portable and on-site, as well as non-destructively screen agricultural commodities sets them apart from traditional analytical or inspection methods that are labor intensive and time-consuming.

Optical and imaging inspection systems have been used increasingly for inspection and evaluation purposes as they can provide rapid, economic, hygienic, consistent and objective assessment. However, difficulties still exist, evident from the relatively slow commercial uptake of these machine vision systems. Even though adequately efficient and accurate algorithms have been generated, processing speeds still fail to meet modern manufacturing requirements, for example. With few exceptions, researches in this field have dealt with trials on a laboratory scale thus the nature of complex biological samples has been neglected. Hence, it needs more focused and detailed study of data mining on agricultural and food matrixes with the aid of such advanced multivariate data analysis as PCA.

5. References

ASTM. (1990). Standard Definitions of Terms and Symbols Relating to Molecular Spectroscopy. American Society for Testing and Materials (ASTM), Vol.14.01, Standard E131-90, ISSN 0066-0531

Abidi, N.; Cabrales, L. & Hequet, E. (2010). Fourier Transform Infrared Spectroscopic Approach to the Study of the Secondary Cell Wall Development in Cotton Fibers. *Cellulose*, Vol.17, pp. 309-320, ISSN 0969-0239

Anderson, J.; Stenzel, S.; Smith, K.; Labus, B.; Rowley, P.; Shoenfeld, S.; Gaul, L.; Ellis, A.; Fyfe, M.; Bangura, H.; Varma, J. & Painter, J. (2002). Multistate Outbreaks of *Salmonella* Serotype Poona Infections Associated with Eating Cantaloupe from Mexico, United States and Canada, 2000-2002. *Morbidity and Mortality Weekly Report*, Vol.51, No.46, pp. 1044-1047, ISSN 0149-2195

Ariana, D.P.; Lu, R. & Guyer, D.E. (2006). Near-Infrared Hyperspectral Reflectance Imaging for Detection of Bruises on Pickling Cucumbers. *Computers and Electronics in Agriculture*, Vol.53, pp. 60-70, ISSN 0168-1699

Burns, D.A. & Ciurczak, E.W. (2001). *Handbook of Near-Infrared Analysis*, Marcel Dekker, Inc., ISBN 0-8247-0534-3, New York, NY, U.S.A.

Chao, K.; Chen, Y-.R. & Chan, D.E. (2003). Analysis of Visible/NIR Spectral Variations of Wholesome, Septicemia, and Cadaver Chicken Samples. *Applied Engineering in Agriculture*, Vol.19, No.4, pp. 453-458, ISSN 0883-8542

Chao, K.; Yang, C.C.; Chen, Y-.R.; Kim, M.S. & Chan, D.E. (2007). Hyperspectral-Multispectral Line-Scan Imaging System for Automated Poultry Carcass Inspection Application for Food Safety. *Poultry Science*, Vol.86, pp. 2450-2460, ISSN 0032-5791

Chen, Y-.R.; Huffman, R.W.; Park, B. & Nguyen, M. (1996). Transportable Spectrophotometer System for On-Line Classification of Poultry Carcasses. *Applied Spectroscopy*, Vol.50, No.7, pp. 910-916, ISSN 0003-7028

Cheng, X.; Chen, Y-. R.; Tao, Y.; Wang, C.Y.; Kim, M.S. & Lefcourt, A.M. (2004). A Novel Integrated PCA and FLD Method on Hyperspectral Image Feature Extraction for Cucumber Chilling Damage Inspection. *Transactions of the ASAE*, Vol.47, No.4, pp. 1313-1320, ISSN 2151-0032

Delwiche, S.R. (2008). High-Speed Bichromatic Inspection of Wheat Kernels for Mold and Color Class Using High-Power Pulsed LEDs. *Sensing and Instrumentation for Food Quality and Safety*, Vol.2, pp. 103-110, ISSN 1932-7587

Delwiche, S.R. & Hareland, G.A. (2004). Detection of Scab-Damaged Hard Red Spring Wheat Kernels by Near-Infrared Reflectance. *Cereal Chemistry*, Vol.81, pp. 643-649, ISSN 0009-0352

Delwiche, S.R.; Kim, M.S. & Dong, Y. (2011a). *Fusarium* Damage Assessment in Wheat Kernels by Vis/NIR Hyperspectral Imaging. *Sensing and Instrumentation for Food Quality and Safety*, Vol.5, pp. 63-71, ISSN 1932-7587

Delwiche, S.R.; Graybosch, R.A.; Amand, P.S. & Bai, G. (2011b). Starch Waxiness in Hexaploid (*Triticum aestivum* L.) by NIR Reflectance Spectroscopy. *Journal of Agricultural and Food Chemistry*, Vol.59, pp. 4002-4008, ISSN 0021-8561

He, L.; Liu, Y.; Lin, M.; Mustapha, A. & Wang, Y. (2008). Detecting Single *Bacillus* Spores by Surface Enhanced Raman Spectroscopy. *Sensing and Instrumentation for Food Quality and Safety*, Vol.2, pp. 247-253, ISSN 1932-7587

Kim, M.S.; Lefcourt, A.M.; Chao, K.; Chen, Y. R.; Kim, I. & Chan, D.E. (2002a). Multispectral Detection of Fecal Contamination on Apples Based on Hyperspectral Imagery: Part I. Application of Visible and Near-Infrared Reflectance Imaging. *Transactions of the ASAE*, Vol.45, No.6, pp. 2027-2037, ISSN 2151-0032

Kim, M.S.; Lefcourt, A.M.; Chen, Y. R.; Kim, I.; Chan, D.E. & Chao, K. (2002b). Multispectral Detection of Fecal Contamination on Apples Based on Hyperspectral Imagery: Part II. Application of Hyperspectral Fluorescence Imaging. *Transactions of the ASAE*, Vol.45, No.6, pp. 2039-2047, ISSN 2151-0032

Kokot, S.; Czarnik-Matusewicz, B. & Ozaki, Y. (2002). Two-Dimensional Correlation Spectroscopy and Principal Component Analysis Studies of Temperature-Dependent IR Spectra of Cotton-Cellulose. *Biopolymers*, Vol.67, pp. 456-469, ISSN 0006-3525

Lawrence, K.C.; Windham, W.R.; Park, B. & Buhr, R.J. (2003). A Hyperspectral Imaging System for Identification of Fecal and Ingesta Contamination on Poultry Carcasses. *Journal of Near Infrared Spectroscopy*, Vol.11, No.4, pp. 269-281, ISSN 0967-0335

Leonard, K.J. & Bushnell, W.R. (2004). *Fusarium Head Blight of Wheat and Barley*, APS Press, ISBN 089-054-302X, Saint Paul, MN, U.S.A.

Liu, Y. & Chen, Y-.R. (2000). Two-Dimensional Correlation Spectroscopy Study of Visible and Near-Infrared Spectral Variations of Chicken Meats in Cold Storage. *Applied Spectroscopy*, Vol.54, No.10, pp. 1458-1470, ISSN 0003-7028

Liu, Y. & Chen, Y-.R. (2003). Analysis of Visible Reflectance Spectra of Stored, Cooked and Diseased Chicken Meats. *Meat Science*, Vol.58, 395-401, ISSN 0309-1740

Liu, Y.; Chen, Y-.R. & Ozaki, Y. (2000). Characterization of Visible Spectral Intensity Variations of Wholesome and Unwholesome Chicken Meats with Two-Dimensional Correlation Spectroscopy. *Applied Spectroscopy*, Vol.54, No.4, pp. 587-594, ISSN 0003-7028

Liu, Y.; Windham, W.R.; Lawrence, K.C. & Park, B. (2003a). Simple Algorithms for the Classification of Visible/Near-Infrared and Hyperspectral Imaging Spectra of Chicken Skins, Feces, and Fecal Contaminated Skins. *Applied Spectroscopy*, Vol.57, No.12, pp. 1609-1612, ISSN 0003-7028

Liu, Y.; Barton, F.E.; Lyon, B.G. & Chen, Y-.R. (2003b). Variations Among Large Spectral Set; Two-Dimensional Correlation Analysis of Loading Spectra from PCA. *Journal of Near Infrared Spectroscopy*, Vol.11, pp. 457-466, ISSN 0967-0335

Liu, Y.; Lyon, B.G.; Windham, W.R.; Lyon, C.E. & Savage, E.M. (2004a). Principal Component Analysis of Physical, Color, and Sensory Characteristics of Cooked Chicken Breasts Deboned at Two, Four, Six, and Twenty-Four Hours Postmortem. *Poultry Science*, Vol.83, pp. 101-108, ISSN 0032-5791

Liu, Y.; Lyon, B.G.; Windham, W.R.; Lyon, C.E. & Savage, E.M. (2004b). Prediction of Physical, Color, and Sensory Characteristics of Broiler Breasts by Visible/Near Infrared Reflectance Spectroscopy. *Poultry Science*, Vol.83, pp. 1467-1474, ISSN 0032-5791

Liu, Y.; Barton, F.E.; Lyon, B.G.; Windham, W.R. & Lyon, C.E. (2004c). Two-Dimensional Correlation Analysis of Visible/Near-Infrared Spectral Intensity Variations of Chicken Breasts with Various Chilled and Frozen Storages. *Journal of Agricultural and Food Chemistry*, Vol.52, pp. 505-510, ISSN 0021-8561

Liu, Y.; Chao, K.; Chen, Y-.R.; Kim, M.S.; Nou, X., Chan, D.E. & Yang, C.C. (2006a). Determination of Key Wavelengths for the Detection of Fecal/Ingesta, Contaminants in Slaughter Plants from Visible and Near-Infrared Spectroscopy. *Journal of Near Infrared Spectroscopy*, Vol.14, pp. 325-331, ISSN 0967-0335

Liu, Y.; Chen, Y-. R.; Wang, C.Y.; Chan, D.E. & Kim, M.S. (2006b). Development of Hyperspectral Imaging Technique for the Detection of Chilling Injury in Cucumbers: Spectral and Image Analysis. *Applied Engineering in Agriculture*, Vol.22, No.1, pp. 101-111, ISSN 0883-8542

Liu, Y.; Chen, Y. R.; Kim, M.S.; Chan, D.E. & Lefcourt, A.M. (2007). Development of Simple Algorithms for the Detection of Fecal Contaminants on Apples from Visible/Near-Infrared Hyperspectral Reflectance Imaging. *Journal of Food Engineering*, Vol.81, pp. 412-418, ISSN 0260-8774

Principal Component Analysis in the Development of Optical and Imaging Spectroscopic Inspections for
Agricultural/Food Safety and Quality

143

Liu, Y.; Delwiche, S.R. & Dong, Y. (2009). Feasibility of FT-Raman Spectroscopy for Rapid Screening for DON Toxin in Ground Wheat and Barley. *Food Additives and Contaminants*, Vol.26, No.10, pp. 1396-1401, ISSN 0265-203X

Liu, Y.; Gamble, G. & Thibodeaux, D. (2010). UV/Visible/Near-Infrared Reflectance Models for the Rapid and Non-Destructive Prediction and Classification of Cotton Color and Physical Indices. *Transactions of the ASAE*, Vol.53, No.4, pp. 1341-1348, ISSN 2151-0032

Liu, Y.; Thibodeaux, D. & Gamble, G. (2011a). Characterization of Attenuated Total Reflection Infrared Spectral Intensity Variations of Immature and Mature Cotton Fibers by Two-Dimensional Correlation Analysis. *Applied Spectroscopy*, Vol.66, No.2 (2012), , ISSN 0003-7028

Liu, Y.; Thibodeaux, D. & Gamble, G. (2011b). Development of FT-IR Spectroscopy in Direct, Non-Destructive, and Rapid Determination of Cotton Fiber Maturity. *Textile Research Journal*, Vol.81, No.15, pp. 1559-1567, ISSN 0040-5175

Lu, R. (2003). Detection of Bruises on Apples Using Near-Infrared Hyperspectral Imaging. *Transactions of the ASAE*, Vol.46, No.2, pp. 1-8, ISSN 2151-0032

Lyon, B.G. & Lyon, C.E. (1991). Research Note: Shear Value Ranges by Instron Warner-Bratzler and Single-Blade Allo-Kramer Devices That Correspond to Sensory Tenderness. *Poultry Science*, Vol.70, pp. 188-191, ISSN 0032-5791

Lyon, B.G. & Lyon, C.E. (1996). Texture Evaluations of Cooked, Diced Broiler Breast Samples by Sensory and Mechanical methods. *Poultry Science*, Vol.75, No.6, pp. 812-819, ISSN 0032-5791

Mahesh, S.; Jayas, D.S.; Paliwal, J. & White, N.D.G. (2011). Identification of Wheat Classes at Different Moisture Levels Using Near-Infrared Hyperspectral Images of Bulk Samples. *Sensing and Instrumentation for Food Quality and Safety*, Vol.5, pp. 1-9, ISSN 1932-7587

Mark, H. & Workman, J. (2007). *Chemometrics in Spectroscopy*, Academic Press, ISBN 0-12-374024-X, Waltham, MA, U.S.A.

Noda, I. (1986). Two-Dimensional Infrared (2D IR) Spectroscopy. *Bulletin of the American Physical Society*, Vol.31, pp. 520, ISSN 0003-0503

Noda, I. (1993). Generalized Two-Dimensional Correlation Method Applicable to Infrared, Raman, and Other Types of Spectroscopy. *Applied Spectroscopy*, Vol.47, No.9, pp. 1329-1336, ISSN 0003-7028

Ozaki, Y.; McClure, W.F. & Christy, A.A. (2007). *Near-Infrared Spectroscopy in Food Science and Technology*, John Wiley & Sons, Inc., ISBN 0-471-67201-7, Hoboken, NJ, U.S.A.

Park, B.; Chen, Y-.R.; Nguyen, M. & Hwang, H. (1996). Characterizing Multispectral Images of Tumorous, Bruised, Skin-Torn, and Wholesome Poultry Carcasses. *Transactions of the ASAE*, Vol.39, No.5, pp. 1933-1941, ISSN 2151-0032

Park, B.; Lawrence, K.C.; Windham, W.R. & Smith, D.P. (2004). Multispectral Imaging System for Fecal and Ingesta Detection on Poultry Carcasses. *Journal of Food Process Engineering*, Vol.27, pp. 311-327, ISSN 1745-4530

Park, B.; Yoon, S.-C.; Windham, W.R.; Lawrence, K.C.; Kim, M.S. & Chao, K. (2011). Line-Scan Hyperspectral Imaging for Real-Time In-Line Poultry Fecal Detection. *Sensing and Instrumentation for Food Quality and Safety*, Vol.5, pp. 25-32, ISSN 1932-7587

USDA. (1996). Pathogen Reduction: Hazard Analysis and Critical Control Point (HACCP) Systems, Final Rule. 9CFR part 304. *Federal Register*, Vol.61, pp. 38805-38989, ISSN 0097-6326

Vargas, A.M.; Kim, M.S.; Tao, Y.; Lefcourt, A.M.; Chen, Y.-R.; Luo, Y.; Song, Y. & Buchanan, R. (2005). Detection of Fecal Contamination on Cantaloupes Using Hyperspectral Fluorescence Imagery. *Journal of Food Science*, Vol.70, No.8, pp. E471-E476, ISSN 1750-3841

Wang, C.Y. (1993). Approaches to Reduce Chilling Injury of Fruits and Vegetables. *Horticultural Reviews*, Vol.15, pp. 63-95, ISBN 0471573388, John Wiley & Sons Inc, Hoboken, NJ, U.S.A

Williams, P. & Norris, K. (2001). *Near-Infrared Technology: In the Agricultural and Food Industries*, American Association of Cereal Chemists, ISBN 1891127241, Saint Paul, MN, U.S.A.

Windham, W.R.; Lawrence, K.C.; Park, B. & Buhr, R.J. (2003). Visible/NIR Spectroscopy for Characterizing Fecal Contamination of Chicken Carcasses. *Transactions of the ASAE*, Vol.46, No.8, pp. 747-751, ISSN 2151-0032

Yang, C.-C.; Chao, K.; Kim, M.S.; Chan, D.E.; Early, H.L. & Bell, M. (2010). Machine Vision System for On-Line Wholesomeness Inspection of Poultry Carcasses. *Poultry Science*, Vol.89, pp. 1252-1264, ISSN 0032-5791

Yang, C.-C.; Kim, M.S.; Kang, S.; Cho, B-.K.; Chao, K.; Lefcourt, A.M. & Chan, D.E. (2012). Red to Far-Red Multispectral Fluorescence Image Fusion for Detection of Fecal Contamination on Apples. *Journal of Food Engineering*, Vol.108, pp. 312-319, ISSN 0260-8774

8

Principal Component Analysis in Industrial Colour Coating Formulations

José M. Medina-Ruiz
University of Minho, Center for Physics
Portugal

1. Introduction

Industrial coatings are functional multilayer thin-films of less than 150 μm that provide a wide range of applications. They are useful to prevent corrosion, to enhance electrical isolation, weather resistance, ultraviolet and infrared protection, decorative purposes etc. Common examples are in everyday commercial products such as in plastics, woods, cosmetics, automotive coatings etc. and in the packaging and security industries (Lewis, 1988; Pfaff & Reynders, 1999; Tracton, 2006). A fundamental issue concerns colour appearance and the consecution of striking visual effects. For this purpose, the introduction of new special-effect colorants has provided a wide range of colour effects that imitate those found for example in Morpho butterflies etc. (Kinoshita et al., 2008). Industrial colour coatings can mimic the surface colour appearance of metals (usually called metallic coatings), the mother pearl effect or pearlescent coatings as well as iridescence or the ability to change the colour as a function of the illumination and the viewing angle (Klein, 2010; Lewis, 1988; McCamy, 1996).

Optical characterization of these industrial coatings demands novel instrumentation and new approaches for precise color formulation in the laboratory and then, for non-destructive testing within the assembled line (Klein, 2010; Völz, 2001). Here I will focus on metallic and pearlescent coatings for automotive paint finishes, one of the largest colour markets in the world. In colour technology, principal-component analysis (PCA) is a standard procedure to uncover the spectral bands of the colorants when they are mixed together (Fairman & Brill, 2004; Kohonen et al., 2006; Liszewski et al., 2010; Ohta, 1973; Tzeng & Berns, 2005), to estimate the number of colorants (Tzeng & Berns, 2005), to simplify the bi-directional reflectance distribution function (BRDF) into few detector positions (Takagi et al., 2005), to indentify the presence of special-effect pigments (Medina, 2008), to evaluate the scattering performance of pigments (Medina & Díaz, 2011), and to develop computer-assisted colour rendering tools for colour styling (Seo et al., 2011). The purpose of this chapter is double. First, I will provide a basic introduction to automotive coatings and paint composition. The standard optical instrumentation will be also presented as well as a basic introduction to colorimetry or how the fundamental attributes of the human colour perception depend on the physical spectra (Wyszecki & Stiles, 1982). And second, I will use PCA in a new and different perspective that consists to examine colour variability for in-line inspection and for pigment identification. There are a considerable number of errors that can modify the final appearance of metallic coatings during the different paint application processes. Accurate colour quality control in the laboratory and in the production line has a fundamental importance for colour recipe correction and fast diagnosis of colour batch production. Here I will show that trial-to-trial variability of reflectance spectra of metallic coatings during the paint application process can be mapped into a stochastic diffusion process. I will use PCA to uncover and to classify the underlying long-range structure of variability and to analyze how this affects to the different types of pigments. An estimation of the Hurst exponent will be given from the eigenvalue spectra (Gao et al., 2003). The Hurst exponent is a measure of the dispersion that has been widely used in the study of anomalous diffusion from the statistical point of view such as in hydrology, finance, etc. (Mandelbrot, 2001). The new approach based on PCA is especially important to better understand the global dependence of colour coating formulations from different car parts manufacturers (Streitberger & Dössel, 2008).

2. Automotive coatings

Typical automotive coatings have four different layers (Streitberger & Dössel, 2008). Fig. 1A represents a schematic representation of the cross-section of a car coating. The electrocoat and the primer surfacer fix the coating to the substrate and prevent against corrosion. After that, the base coat or binder constitutes the principal element and provides the fundamental characteristics related with colour, elasticity, brightness, dispersion and chemical resistance. Finally, the outermost layer is the clear coat or the transparent lacquer that protects the binder from the exterior. Different types of pigments are included in the binder. Their selection depends on the nature of the resin (polyesters, acrylics, etc.) and the paint application technology such as solvent-based, water-based (environmentally friendly) or powder coatings.

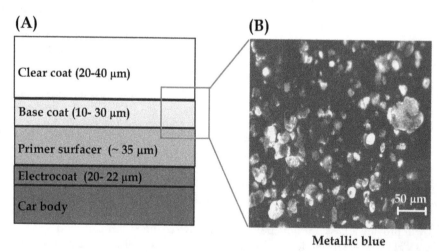

Fig. 1. (A) Schematic representation of the cross-section of a typical car coating and related layer thickness (in µm). (B) Optical micrograph of a metallic blue coating containing aluminium flakes and conventional absorption blue pigments in the base coat. Bright field illumination (20x).

The principal functions of pigments are to obtain a specific opacity or hiding power, to give remarkable colour effects, to provide stability against direct sunlight exposition and to provide resistance against different factors that affects corrosion. Some pigments are organic in nature and others are inorganic. Regarding special-effect pigments, they are metal-like (e.g. aluminium), pearl-like and interference flakes. They have, on average, a greater size (typically in μm) in comparison with conventional absorption pigments (in nm) and are often used to produce unusual colour variations in relation to the viewing angle (Klein, 2010; Lewis, 1988; Pfaff & Reynders, 1999; Tracton, 2006). Fig. 1B shows an optical micrograph of a typical metallic blue coating containing lenticular aluminium flakes covered by organic blue (indanthrone) pigment nanoparticles in the base coat. Chemical pigments scatter the light in all directions or diffusive scattering. Further, they selectively absorb the incident radiation and can be formulated in accordance with the laws of subtractive colour mixing. Aluminium flakes, however, act like reflectors and scatter the light at the corners. In a different type of pigments, bright iridescent colours are often produced by thin-film interference pigments that obey the laws of additive colour mixing (Klein, 2010; Lewis, 1988; Pfaff & Reynders, 1999; Tracton, 2006). Very popular interference pigments in automotive coatings are those containing mica flakes coated by different metal oxides such as titanium dioxide (TiO_2) (Klein, 2010; Lewis, 1988; Pfaff & Reynders, 1999; Tracton, 2006). For a detail description of the deposition of pigments in the base coat in car coatings see (Streitberger & Dössel, 2008).

Fig. 2A shows an optical micrograph of the metallic green coating containing yellowish-greenish interference pigments as indicated in example selected by the red square. In a perfect regular multilayer stack as the ideal case represented in Fig.2B, light at the interface of a tiny mica flake is partially reflected and partially transmitted producing a phase shift from a low-to-high refractive index media. Angle-dependent colour effects can arise from constructive interference (usually called goniochromism) (McCamy, 1996). In accordance with the Bragg condition, the reflected light is wavelength dependent and obeys the following equation (Hecht & Zajac, 1974; Klein, 2010; Lewis, 1988):

$$2\,\text{nd}\cos(\alpha) = \left(m + \frac{1}{2} \right)\lambda \qquad (1)$$

where m is an integer and λ is the wavelength. The Bragg condition is also dependent on the refractive index of the material n, the thickness d, as well as the angle of refraction α between the illumination and the normal to plane of the sample. By selecting the appropriate thickness, tunable interference colours can de designed. The reflected rays are parallel to each other and the perception of colour (in our particular case yellowish) depends on the viewing angle (Klein, 2010; Lewis, 1988; Pfaff & Reynders, 1999). The transmitted light through mica pigments corresponds to the complementary colour (i.e. bluish) and can interact with the substrate (black or white) (Klein, 2010; Lewis, 1988; Pfaff & Reynders, 1999). In real mica-based pigments thickness is not a constant parameter and different colours can be observed inside the same flake as revealed in Fig. 2A. The control of the spatial dispersion of special-effect pigments is an important factor to obtain the desired colour effect. Both aluminium and interference flakes are, on average, oriented parallel to the substrate (Klein, 2010; Lewis, 1988; Streitberger & Dössel, 2008). The flake orientation distribution can follow a Gaussian-type function with a long right tail (Kirchner &

Houweling, 2009; Sung et al., 2002). An important topic in car coatings is the concept of "colour harmony" or no perceptual colour differences between the different external parts of the car. Fig.3 shows an example of colour harmony.

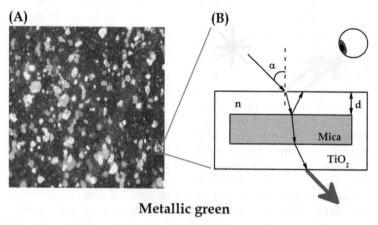

Metallic green

Fig. 2. (A) Optical micrograph of a metallic green coating containing greenish-yellowish interference pigments, aluminium flakes and opaque green absorption pigments. Darkfield illumination (20x). (B) Schematic representation of the cross-section of a mica flake coated by titanium dioxide (TiO_2) with thickness d and refraction index n.

Fig. 3. Example of colour harmony in a silver-like coating containing aluminium flakes. Current colour designs must achieve no perceptible colour differences between the different external parts that integrate the car such as the car body and the fenders.

The analysis of colour harmony is a complex issue because bumpers, fenders, wings etc. are often provided by different manufacturers, each of them with similar paint procedures, and the car parts should be assembled all together in a specific automotive production line (Streitberger & Dössel, 2008). I will show below that PCA can be applied to examine the intricate variability of the paint application processes during colour batch production and, thus, it can be used to study colour harmony.

3. Spectroscopic instrumentation

Spectrophotometer-based systems provide an estimation of the spectral reflectance distribution function, which describes the ratio of the reflected radiance to the incident irradiance from a surface for a specific illumination and detection positions (Klein, 2010; Völz, 2001; Wyszecki & Stiles, 1982). Actually the CIE (Commission Internationale de l'Éclairage, International Commission on Illumination), promotes and regulates the standards conditions for reflectance and colour measurements. The measurement of the spectral reflectance and the colorimetric analysis of thin-films based on conventional chemical pigments and dyes have greatly simplified because matte and glossy surfaces have residual or no angular dependence. This is not the case of automotive coatings containing special-effect pigments and they require the estimation of the spectral reflectance function in the hemisphere centre at the sample or the BRDF.

The BRDF provides the spectral reflectance function of a surface measured in spherical coordinates using directional illumination. The complete characterization of the BRDF often assumes many positions or degrees of freedom (Baba & Arai, 2003). This complicates the acquisition of reflectance spectra with a large number of readings at specific illumination and detection positions and their subsequent colorimetric evaluation in automotive coatings, a problem that has not been solved yet. Few standards of gonioappearance have appeared in the last years and portable multi-angle spectrophotometers are now using for testing and colour quality control. Fig. 4 shows a schematic representation of the detection geometries recommended for metallic coatings.

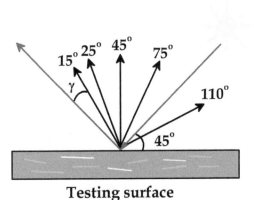

Testing surface

Fig. 4. Schematic representation of the cross-section of a metallic coating and the directional illumination at 45° and the 5 angular detection configuration at the aspecular angle γ of 15°, 25°, 45°, 75° and 110°.

For a fixed illumination angle (usually 45° off from the normal of the measured surface), a set of predetermined detector angles is often defined from the specular reflection. DIN (The Deutsches Institut für Normung) recommends the use of three angles of 25°, 45°, 75°, and optionally 110° from the specular (usually called the "aspecular angle" γ) (DIN 6175-2, 2001), whereas the ASTM (American Society for Testing Materials), prescribes three angles of 15°, 45°, and 110° (ASTM E2175-01, 2001). However, previous studies in pearlescent coatings have probed that these measurement geometries are not enough due to the angular dependence of the interference pigments on the illuminant position (see Eq. 1) (Cramer, 2002; Nadal & Early, 2004). New industrial rules have been proposed (ASTM E2539-08, 2008) and novel scientific instrumentation is coming soon. Here I will restrict the PCA of reflectances using simultaneously those geometries compatible with the DIN (DIN 6175-2, 2001) and the ASTM (ASTM E2175-01, 2001).

4. Colorimetry: Basic concepts and definitions

The goal of colorimetry is to provide a simplified analysis of the illuminants and colouring surfaces using a three-dimensional colour space, which is a theoretical construction for classification and precise representation of colour properties (Wyszecki & Stiles, 1982). Given a directional illumination source characterized by the energy distribution $S(\lambda)$, a testing surface with spectral reflectance $R(\lambda)$ and a observer (e.g. the human eye) located at a fixed detection position, colour properties of the surface can be labelled in the colour space using the colour coordinates. By the trichromatic principle (Wyszecki & Stiles, 1982), the colour coordinates can be defined in the CIE standard colorimetric system by the tristimulus values X, Y and Z (Klein, 2010; Völz, 2001; Wyszecki & Stiles, 1982):

$$X = k \int S(\lambda)R(\lambda)\bar{x}_{10}(\lambda)d\lambda$$
$$Y = k \int S(\lambda)R(\lambda)\bar{y}_{10}(\lambda)d\lambda \qquad (2)$$
$$Z = k \int S(\lambda)R(\lambda)\bar{z}_{10}(\lambda)d\lambda$$

Where $\bar{x}_{10}(\lambda), \bar{y}_{10}(\lambda), \bar{z}_{10}(\lambda)$ are the CIE-1964 colour matching functions. The CIE-1964 colour matching functions define the spectral sensitivity of human eye for a standard observer that subtends a 10° field of view. Spectral integration in Eq. 2 is often done between 400-700nm. The constant k defines a normalization factor so that for a perfect white reflectance surface of $R(\lambda) = 1$, the tristimulus value Y, which is related with the luminance content, equals to 100 for every illuminant (Klein, 2010; Völz, 2001; Wyszecki & Stiles, 1982):

$$k = \frac{100}{\int S(\lambda)\bar{y}_{10}(\lambda)d\lambda} \qquad (3)$$

Fig.5A represents the relative energy distribution of the illuminant D65, which imitates daylight conditions, and the standard incandescent illuminant A. Fig. 5B represents the spectral shape of the CIE-1964 colour matching functions.

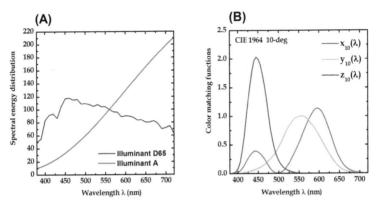

Fig. 5. (A) Relative energy distribution of the standard illuminants D65 and A recommended by the CIE. (B) CIE-1964 colour matching functions for the 10° standard observer.

For pigmented thin-films as those in automotive coatings, it is more convenient to provide a transformation from the CIE-1964 colour space to the CIELAB perceptual colour space (Klein, 2010; Völz, 2001). In the CIELAB colour space, the new orthogonal colour coordinates are L* (the luminance axis), related with the brightness sensation (i.e. the white and black contribution) and the chromaticity coordinates a* (red-green axis) and b* (blue-yellow axis) that defines the chromatic plane. From the tristimulus values defined in Eqs. 2-3, the CIELAB coordinates can be calculated as follows (Klein, 2010; Völz, 2001; Wyszecki & Stiles, 1982):

$$\forall \left(\frac{X}{X_n}\right), \left(\frac{Y}{Y_n}\right), \left(\frac{Z}{Z_n}\right) > 0.008856 \Rightarrow$$

$$\Rightarrow \begin{cases} X^* = \sqrt[3]{\dfrac{X}{X_n}} \\[2mm] Y^* = \sqrt[3]{\dfrac{Y}{Y_n}} \\[2mm] Z^* = \sqrt[3]{\dfrac{Z}{Z_n}} \end{cases} \tag{4}$$

where X_n, Y_n and Z_n are the CIE tristimulus values of the reference illuminant used (Y_n equals to 100 for all the standard illuminants, see Eqs. 2-3):

$$\forall \left(\frac{X}{X_n}\right), \left(\frac{Y}{Y_n}\right), \left(\frac{Z}{Z_n}\right) \leq 0.008856 \Rightarrow$$

$$\Rightarrow \begin{cases} X^* = 7.787\dfrac{X}{X_n} + 0.138 \\[2mm] Y^* = 7.787\dfrac{Y}{Y_n} + 0.138 \\[2mm] Z^* = 7.787\dfrac{Z}{Z_n} + 0.138, \end{cases} \tag{5}$$

$$L^* = 116\,Y^* - 16$$
$$a^* = 500\left(X^* - Y^*\right) \tag{6}$$
$$b^* = 200\left(Y^* - Z^*\right)$$

Positive values of a* and b* represents reddish and yellowish, respectively whereas negative values indicate greenish and bluish, respectively (Klein, 2010; Völz, 2001; Wyszecki & Stiles, 1982). Colour coordinates a* and b* defined in Eq.6 are reminiscent of the chromatic-opponent physiological organization of ganglion cells at the retina (Dacey, 2000). As an example, Fig.6A shows the reflectance spectra measured with a multi-angle spectrophotometer (400- 700 nm at 10 nm steps) for a typical green metallic coating containing aluminium, opaque green absorption nanopigments and mica-based interference pigments at the aspecular angles of 15°, 25°, 45°, 75° and 110°.

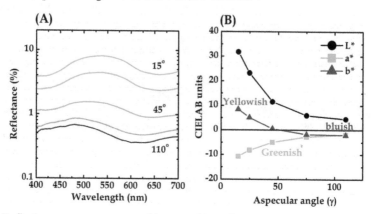

Fig. 6. (A) Reflectance spectra measured by a multi-angle spectrophotometer (X-Rite MA68-II) at the aspecular angle γ of 15°, 25°, 45°, 75° and 110°. (B) CIE L*, a* and b* colour coordinates calculated at different the aspecular angles. The reference illuminant was D65.

Because both aluminium and mica flakes are mainly oriented parallel to the substrate, bright interference colours are clearly manifested near the specular reflection at the aspecular angles γ of 15° and 25°, whereas reflection from diffusive multiple-light scattering and pigment absorption dominate far from the specular such as the aspecular angles γ of 75° and 110° (Klein, 2010). In Fig. 6A maximum of reflectance shifts from greenish-yellowish (540 nm) near the specular at 15° to greenish-bluish (480 nm) far from the specular at 110°. Fig. 6B shows the CIELAB colour coordinates L*, a*, and b* as a function of the aspecular angle γ under the reference illuminant D65. Lightness values change from 31.8 CIELAB units at γ= 15° to 4.48 CIELAB units at γ= 110°. Direct visual observation with the naked eyes of the metallic green panel in a light booth and using a D65 daylight lamp validate the above colorimetric description.

5. Principal component analysis of reflectances

PCA is one common way to decorrelate reflectance spectra into few components that maximizes the variance accounted for (Fairman & Brill, 2004; Ohta, 1973; Tzeng & Berns, 2005).

For this purpose, it is possible to treat the reflectance spectra of metallic coatings as vectors (as much dimensions as wavelengths) in the vector space of square-integrable functions L^2. For example, given a reflectance dataset composed by N different reflectance elements between 400- 700nm at 10nm steps (i.e. 31 dimensional vectors), it is possible to establish a linear decomposition of each spectral reflectance function $R_j(\lambda)$ around the mean reflectance of the database $\bar{R}(\lambda)$ (Fairman & Brill, 2004; Ohta, 1973; Tzeng & Berns, 2005):

$$\forall j = 1, 2, \ldots, N \Rightarrow$$
$$\Rightarrow R_j(\lambda) = \bar{R}(\lambda) + \sum_{i=1}^{31} b_i Z_i(\lambda) \tag{7}$$

where the coefficients b_i are the coordinates in the new reference system or the principal components and $Z_i(\lambda)$ are the new bases or eigenvectors in the L^2 vector space. The first basis function $Z_1(\lambda)$ is uncorrelated with the second basis function $Z_2(\lambda)$ and its associated eigenvalue explains most of the variance and so on. Therefore, PCA of reflectance spectra of metallic coatings at each aspecular angle separately can uncover the spectral signature of the basis functions and permit to compare the bases across different aspecular angles for pigment identification (Medina, 2008).

5.1 Anomalous diffusion assessed by principal component analysis

Variability in the reflectance spectra and, thus, in the colour batch production of bumpers, fenders, wings, etc. can be mapped into a one-dimensional random walk model. The classical example of random walk is the irregular Brownian molecular motion of a pollen grain suspended in a liquid. Fig. 7A simulates a hypothetical Brownian diffusion process for

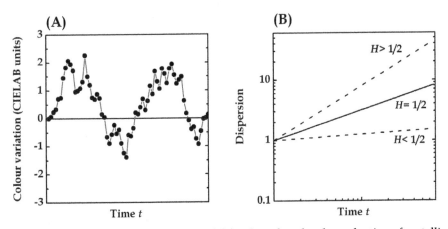

Fig. 7. (A) Illustration of a random walk model for the colour batch production of metallic coatings. Black dots in the time series correspond to the CIELAB colour coordinates measured in each painted panel relative to the master panel within the production line. (B) Double logarithmic plot of the dispersion as a function of time for different values of the Hurst exponent H. Solid and dashed lines correspond to the standard Brownian and fractional Brownian motion processes, respectively.

colour production of metallic coatings. Here the position of the random walker at a single step refers to each painted piece in chronological order and is represented by the CIELAB colour coordinates relative to the reference master panel, i.e. ΔL^*, Δa^*, Δb^* and can assume positive and negative values (Klein, 2010; Völz, 2001; Wyszecki & Stiles, 1982).

In the proposed model, the mean square displacement of the colour differences can grow with time by a power law, t^{2H}, where H is the Hurst exponent and gives a measure of dispersion. A standard Brownian process has H= 0.5 and defines the reference baseline (i.e. no correlation) to indentify the presence of anomalous diffusion. The generalization of the standard Brownian motion is called fractional Brownian motion with the Hurst exponent taking values between zero and unity (Mandelbrot & Van Ness, 1968). When H> 0.5 colour coordinates differences are positively correlated suggesting the presence of persistent deviations, i.e., an increase in the colour coordinates differences are more probably to be followed by another increase later in the scale of time. When H< 0.5 colour coordinates differences are negatively correlated suggesting the presence of antipersistent deviations. That is, after an increase in the colour coordinates differences of the painted pieces relative to the master panel, it is more probably to obtain a decrease in the colour differences later in the scale of time. It has been proved that for a fractional Brownian motion process, the eigenvalue spectrum derived from PCA decays as a power law with exponent β. This exponent is related with the Hurst exponent (Gao et al., 2003):

$$\beta = 2H + 1 \tag{8}$$

The exponent β in Eq. 8 can also take values lower than unity (i.e. H= 0) and higher than three (i.e. H= 1). In the former case, it may be related with fractional Gaussian processes defined as the derivative of the fractional Brownian motion (Mandelbrot, 2001; Mandelbrot & Van Ness, 1968). A different definition of the exponent β may be required and this issue remains to be determined. For β>3, they correspond to anomalous diffusion that grows faster than any traditional fractional Brownian motion process (Medina & Díaz, 2011). Therefore, a double logarithmic plot of the eigenvalues as a function of the basis function order is a simple and valuable method to uncover the existence of power laws at different aspecular angles and, thus, to identify the presence of anomalous diffusion in metallic coatings using the model in Eq.8 as the reference framework (see Fig. 7B) (Gao et al., 2003; Medina & Díaz, 2011).

6. Example of principal component analysis in metallic coatings

6.1 Colorimetric description

Figs. 8A and 8B represent the reflectance spectra and the CIE L* a* b* colour coordinates of a typical blue metallic coating, respectively. Data correspond to the master panel at the aspecular angles of 15°, 25°, 45°, 75° and 110° and were taken with a conventional multi-angle spectrophotometer (X-Rite MA68 II). White and zero calibration were done in regular time sequences using a white ceramic tile and a black trap, respectively.

Near the specular reflection at 15° and 25°, maximum of reflectance is found at 460 nm indicating a greenish-bluish colour whereas far from the specular the peak is reduced to almost the same reflectance value as in 540 nm and 640 nm and promotes the perception of bluish-reddish. Luminance values change from 30 CIELAB units at 15° to 3.8 CIELAB units at 110°. Fig. 9 shows in the CIELAB colour space a three-dimensional representation of a test colour batch production (blue spheres).

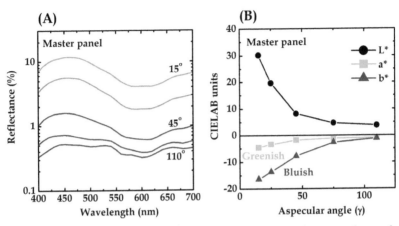

Fig. 8. (A) Reflectance spectra of a typical blue metallic coating at the aspecular angle γ of 15°, 25°, 45°, 75° and 110°. (B) CIE L*, a* and b* colour coordinates calculated at different the aspecular angles. The reference illuminant was D65. Data indicate the master panel.

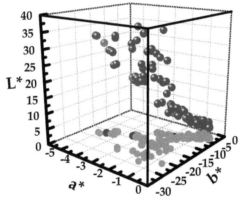

Fig. 9. Three-dimensional representation of the CIE L*, a* and b* colour coordinates corresponding to the blue metallic panels at the aspecular angle γ of 15°, 25°, 45°, 75° and 110°. Blue spheres correspond to the different painted panels whereas red spheres indicate the colour coordinates of the master panel. Dark cyan and dark red solid circles represent the projection in the a*b* plane, respectively. The reference illuminant was D65.

They consist of 21 successive painted panels and are representative of original equipment manufacturers in solventborne coatings (Streitberger & Dössel, 2008). They were painted in the production line of a car supplier in Europe using a colour recipe to replicate the colour properties of the master panel in Fig. 8. They have aluminium flakes, opaque blue and carbon black absorption pigment nanoparticles and also mica-based interference flakes in a complex binder containing polyester, melamine and wax. The lacquer was transparent and was the same in all the 21 painted pieces. Reflectance spectra were measured at the end of the painted process at the 5 aspecular angles of 15°, 25°, 45°, 75° and 110°. Therefore, the reflectance database consists of a total of 105 reflectances between 400- 700nm at 10nm steps. Then, they were converted to the CIELAB colour space using the D65 illuminant. The projection in the a*b* plane is also displayed in Fig.9. Red spheres represent the colour coordinates of the master panel at different aspecular angles from Fig. 8B. The results indicate that the painted blue metallic panels occupy a characteristic colour volume and expand a specific colour map in the a*b* plane around the reference master panel. In the colour batch production, the range (maximum – minimum) of luminance variations over the 5 aspecular angles was 35 CIELAB units whereas in the red-green axis a* and in the blue-yellow axis b* the range was 4.6 and 22 CIELAB units, respectively.

6.2 Basis functions

PCA of reflectances was done over the 21 painted panels at each aspecular angle separately. Fig. 10 represents the first six basis functions near the specular reflection at 15° (cyan solid lines) and far from the specular reflection at 110° (blue solid lines). The first eigenvector or basis function has a peak around 410- 430 nm and its shape remains almost invariant as a function of the aspecular angle. It accounts for more than 98% of the total variance at the aspecular angle of 15° but only around 86% of total variance at 110°. It can be interpreted as the result of the aluminium flakes, which has a flat reflectance spectrum except at 400-420 nm, covered by opaque absorption blue pigment nanoparticles in the binder (Baba & Arai, 2003; Medina, 2008), in the same way as the spatial deposition illustrated in Fig 1B. That means that most of the variability associated with the blue metallic coatings corresponds to lightness and bluish changes as verified in Fig. 8B and Fig. 9.

The second and third basis functions are peaked a different wavelengths and represents more than 99% of the total variance accounted for. Their shape look similar and present a wavelength shift from 480 nm at 15° to 410 nm at 110° in the second basis function and from 420 nm at 15° to 480 nm at 110° in the third basis function. The effect of the aspecular angle on the basis functions can be interpreted as the presence of interference pigments within the base coat (Medina, 2008) (see also Eq. 1 and Fig. 2B). The next basis functions correspond to different corrections and are more difficult to interpret. This issue will be discussed further in the next section. In colour technology, a common procedure is to select only those basis functions that achieve 99% of the total variance accounted for or in some cases even more (Tzeng & Berns, 2005). For example, in Fig. 10 it is required six basis function to represent more than 99.5% of the total variability of the reflectance spectra. I will show below that this statistical criterion is not adequate to examine the presence of diffusion process.

6.3 Eigenvalue spectra

Fig. 11 represents in a double logarithmic plot, the eigenvalue spectra of the blue metallic coatings from PCA at each aspecular angle. Only those eigenvalues at 15°, 45° and 110° are displayed (the eigenvalue spectra for the aspecular angle of 25° and 75° are in intermediate positions). The results confirm that the eigenvalues decrease as the basis function order increases. Power laws were fitted using a least square procedure and the goodness-of-fit was assessed by the coefficient of regression R-square or R^2 (Press et al., 1992). Far from the specular reflection at 110° (blue triangles), from the third to the nineteenth eigenvalue a single power law regime is found representing 2.14% of the total variance account for. The coefficient of regression was $R^2 = 0.99$ and the estimated exponent was $\beta = -3.9$. This suggests the presence of persistence correlations that grows faster than any fractional Brownian diffusion processes. They can be associated with the trial-to-trial variability in the spatial dispersion of absorption pigments within the base coat.

Metallic blue

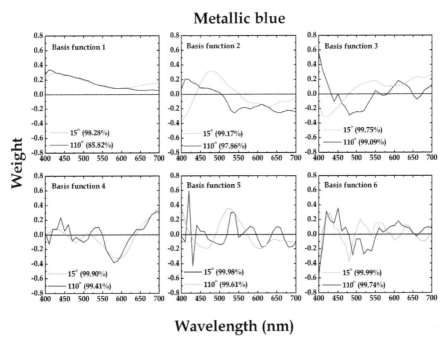

Fig. 10. PCA of reflectances from blue metallic coatings. Panel shows the first six basis functions. The percentage in parentheses indicates the cumulative variance account for from near-specular reflection at 15° (cyan lines) to far from the specular at 110° (blue lines).

At 110° there is an upper cut-off that affects the first two eigenvalues and may reflect a finite-size effect from a small quantity of aluminium flakes and from the interference flakes that have a different orientation in comparison with the substrate. At all the aspecular angles, there is also a second limiting physical process in the eigenvalue spectrum that produces a lower cut-off. Interestingly, when we move from far to the specular reflection to an intermediate position such as at 45° (grey squares), the first six basis function deviates from the above power law regime at 110°, this suggesting that more eigenvalues are affected by the presence aluminium and interference pigments oriented quasi-parallel to the substrate. Near the specular reflection at 15°, reflectance data contains most of the contribution from the aluminium and interference flakes oriented parallel to the substrate. From the second to the fifth eigenvalue a new power law regime is found ($R^2 = 0.86$). The power law represents 1.70% of the total variance accounted and has an exponent $\beta = -1.9$. Since the first two basis functions in Fig. 10 are associated with the presence of mica-based flakes, the eigenvalue analysis suggest that trial-to-trial reflectance variability from interference pigments across different painted panels increase as a function of time following a Brownian diffusion process or equivalently, showing neither persistent nor antipersitent correlations (a Hurst exponent very close to H= 0.5). The second power-law regime at 15° goes from the seventh to the nineteenth eigenvalue (R^2= 0.99). This power law has a slope β= -3.8, and, thus, very similar to the contribution expected at 110° from absorption pigments but only represents less than 0.01% of the total variance accounted for.

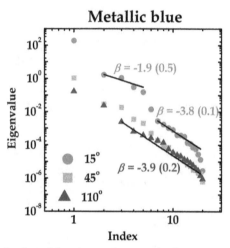

Fig. 11. Double logarithmic plot of the eigenvalue spectra from PCA of reflectances of the blue metallic coatings. Solid dark cyan circles, grey squares and blue triangles denote the eigenvalue spectrum at the aspecular angle γ of 15° (near-specular), 45° (intermediate) and 110° (far from the specular), respectively. Black solid lines indicate the adjusted power laws at different regimes. The slope β denote the scaling exponent. Numbers in parentheses indicate their associate standard errors.

7. Conclusions

Principal component analysis of reflectance spectra is a powerful method to determine the possible sources of variability in surface colour appearance of industrial coatings. In comparison with the common statistical criterion used in colour technology that recommends to retain only few basis functions (Tzeng & Berns, 2005), here a new analysis is introduced to examine the eigenvalue spectrum by power laws (Gao et al., 2003). Therefore, it was investigated the trial-to-trial variability in the colour batch production of metallic coatings in terms of fractional Brownian diffusion process and their correlations. This is important in colour recipe formulation and offers a novel approach to study colour harmony between different paint manufacturers. The PCA made in this chapter have dealt with the reflectance spectra collected at different aspecular angles and is representative of automotive coatings containing special-effect pigments (Streitberger & Dössel, 2008). For the blue metallic coatings, it was concluded that the trial-to-trial variability of mica-based pigments across different painted panels is not correlated and follows a Brownian diffusion process with a Hurst exponent compatible with H= 0.5. However, persistent trial-to-trial correlations with the same exponent at different aspecular angles were found, this suggesting a contribution from the deposition of absorption pigments in the base coat (see Fig. 11). This may be related with the use solventborne coatings and how the pigments are dispersed in the binder (Streitberger & Dössel, 2008). Therefore, it would desirable to comparable these results with similar paint application processes as well as in waterborne and powder coatings. Finally, the methods presented in this chapter are not restricted to car coatings and can be applied in the study of different industrial colour process and coloured thin-films.

8. Acknowledgments

This work was supported by the European Regional Development Fund through Programa Operacional Factores de Competitividade—COMPETE (FCOMP-01-0124-FEDER-014588), by the National Portuguese Funds through the Fundação para a Ciência e Tecnologia—FCT (PTDC/CTM-MET/113352/2009), and by the Center for Physics, University of Minho, Portugal.

9. References

ASTM E2175-01 (2001). Standard Practice for Specifying the Geometry of Multiangle Spectrophotometers. (American Society for Testing and Materials).

ASTM E2539-08 (2008). Standard Practice for Multiangle Color Measurement of Interference Pigments. (American Society for Testing and Materials).

Baba, G. & Arai, H. (2003). Gonio-spectrophotometry of metal-flake and pearl-mica pigmented paint surfaces. In Fourth Oxford Conference on Spectroscopy, A. Springsteen, and M. Pointer, eds. (Proceedings of the Society of Photo-Optical Instrumentation Engineers (SPIE)), pp. 79-86.

Cramer, W.R. (2002). Examples of interference and the color pigment mixtures green with red and red with green. *Color Research and Application* 27, pp. 276-281, ISSN 0361-2317.

Dacey, D.M. (2000). Parallel pathways for spectral coding in primate retina. *Annual Review of Neuroscience* 23, pp. 743-775, ISSN 0147-006X.

DIN 6175-2 (2001). Farbtoleranzen für Automobillackierungen - Teil 2: Effektlackierungen. (Deutsches Institut für Normung e.V.).

Fairman, H.S. & Brill, M.H. (2004). The principal components of reflectances. *Color Research and Application* 29, pp. 104-110, ISSN 0361-2317.

Gao, J.B., Cao, Y.H. & Lee, J.M. (2003). Principal component analysis of 1/f(alpha) noise. *Physics Letters A* 314, pp. 392-400, 0375-9601.

Hecht, E. & Zajac, A. (1974). *Optics* Addison-Wesley Publishing Company, Inc, ISBN 978-0805385663.

Kinoshita, S., Yoshioka, S. & Miyazaki, J. (2008). Physics of structural colors. *Reports on Progress in Physics* 71, pp. 30, ISSN 0034-4885.

Kirchner, E. & Houweling, J. (2009). Measuring flake orientation for metallic coatings. *Progress in Organic Coatings* 64, pp. 287-293, 0300-9440.

Klein, G.A. (2010). *Industrial Color Physics* Springer Science+Business Media LLC, ISBN 978-1-4419-1196-4, New York.

Kohonen, O., Parkkinen, J. & Jaaskelainen, T. (2006). Databases for spectral color science. *Color Research and Application* 31, pp. 381-390, ISSN 0361-2317.

Lewis, P.A. (1988). *Pigment handbook, properties and economics*, Vol 1, 2 edn John Willey & Sons, ISBN 0-471-82833-5, New York.

Liszewski, E.A., Lewis, S.W., Siegel, J.A. & Goodpaster, J.V. (2010). Characterization of Automotive Paint Clear Coats by Ultraviolet Absorption Microspectrophotometry with Subsequent Chemometric Analysis. *Applied Spectroscopy* 64, pp. 1122-1125, ISSN 0003-7028.

Mandelbrot, B.B. (2001). *Gaussian Self-affinity and Fractals* Springer-Verlag, ISBN 0387989935, New York.

Mandelbrot, B.B. & Van Ness, J.W. (1968). Fractional brownian motions, fractal noises and applications. *Siam Review* 10, pp. 422-437, ISSN 0036-1445.

McCamy, C.S. (1996). Observation and measurement of the appearance of metallic materials .1. Macro appearance. *Color Research and Application* 21, pp. 292-304, ISSN 0361-2317.

Medina, J.M. (2008). Linear basis for metallic and iridescent colors. *Applied Optics* 47, pp. 5644-5653, ISSN 0003-6935.

Medina, J.M. & Díaz, J.A. (2011). Scattering characterization of nanopigments in metallic coatings using hyperspectral optical imaging. *Applied Optics* 50, pp. G47-G55, ISSN 0003-6935.

Nadal, M.E. & Early, E.A. (2004). Color measurements for pearlescent coatings. *Color Research and Application* 29, pp. 38-42, ISSN 0361-2317.

Ohta, N. (1973). Estimating absorption-bands of component dyes by means of principal component analysis. *Analytical Chemistry* 45, pp. 553-557, ISSN 0003-2700.

Pfaff, G. & Reynders, P. (1999). Angle-dependent optical effects deriving from submicron structures of films and pigments. *Chemical Reviews* 99, pp. 1963-1981, ISSN 0009-2665.

Press, W., Teukolsky, S., Vetterling, W. & Flannery, B. (1992). *Numerical recipes in C* Cambridge University Press, 0-521-43108-5, New York.

Seo, M.K., Kim, K.Y., Kim, D.B. & Lee, K.H. (2011). Efficient representation of bidirectional reflectance distribution functions for metallic paints considering manufacturing parameters. *Optical Engineering* 50, ISSN 0091-3286.

Streitberger, H.-J. & Dössel, K.-F. (2008). *Automotive Paints and Coatings*, Second edn Wiley-VCH Verlag GmbH & Co. KGaA, ISBN 978-3527309719.

Sung, L.P., Nadal, M.E., McKnight, M.E., Marx, E. & Laurenti, B. (2002). Optical reflectance of metallic coatings: Effect of aluminum flake orientation. *Journal of Coatings Technology* 74, pp. 55-63, ISSN 0361-8773.

Takagi, A., Watanabe, A. & Baba, G. (2005). Prediction of spectral reflectance factor distribution of automotive paint finishes. *Color Research and Application* 30, pp. 275-282, ISSN 0361-2317.

Tracton, A.A. (2006). *Coatings Technology Handbook*, Third edn CRC press, Taylor & Francis Group, LLC, ISBN 1-574-446495, Boca Raton, Florida.

Tzeng, D.Y. & Berns, R.S. (2005). A review of principal component analysis and its applications to color technology. *Color Research and Application* 30, pp. 84-98, ISSN 0361-2317.

Völz, H.G. (2001). *Industrial color testing, fundamentals and techniques*, 2 edn Wiley VCH Verlag GmbH & Co. KGaA, ISBN 3-527-30436-3.

Wyszecki, G. & Stiles, W.S. (1982). *Color science: concepts and methods, quantitative data and formulae*, 2 edn John Wiley & Sons, ISBN 0-471-021067, New York.

9

Improving the Knowledge of Climatic Variability Patterns Using Spatio-Temporal Principal Component Analysis

Sílvia Antunes[1], Oliveira Pires[2] and Alfredo Rocha[3]
[1]Institute of Meteorology
[2]Meteorological and Geophysical Portuguese Association
[3]CESAM & Department of Physics of University of Aveiro
Portugal

1. Introduction

We may define climate as a statistical synthesis of the weather conditions at a given place or region during a period of time. Climate is different from place to place and also changes with time. For instance, the climate is different in the British Islands, in Central Europe, in the Iberian Peninsula, in the Sahara or in the Amazons. The climate changes from Winter to Summer and from daytime to nigh time.

The climate also shows variations over long periods of time. There are strong evidences of the existence in the past times of glacial and interglacial periods. A brief synthesis on this subject can be seen, for instance, in Duplessy and Morel (1990). There are also evidences of changes in climate over shorter time scales. Some of them are not detectable from direct instrument measures, which just began in a systematic way after the Second World War (Peixoto & Oort, 1992). Considerable effort has been made during the last decade to reconstruct global or northern hemispheric temperatures for the past in a long term perspective (Ljungqvist, 2010), referring for example: von Storch et al. (2004), Wanner et al. (2008), Osborn & Briffa (2006), Mann & Jones (2003), Lee et al. (2008) and Jones et al. (2009).

Besides these changes in time, the climate shows year to year variability.

For quantitative studies of the climate in the near past or present times series of measurements of the climatic elements such as air temperature, atmospheric pressure, precipitation amount, etc., are analysed.

Climatic series are composed in such a way that they filter the seasonal and daily variability. For instance, it can be used climatic chronological series of values of mean air temperature in every January year after year, series of total annual precipitation amount year after year, series of mean atmospheric pressure during every spring year after year, etc.

The climatic series formed in this way do not show seasonality or daily variability and have a statistical behaviour that is, in a first approximation, similar to realizations of white noise or red noise stochastic processes. The principal aim of the statistical climatology analysis is to detect in the series properties that significantly differ that what should be expected from simple realizations of stationary white or red noise. These properties are called "signals" and the remainder part of the series is called the "noise". Typically climatic series have low signal to noise ratios or, in other words, the fraction of variance explained by the signal (or signals) is generally low when compared to the variance of the remainder noise.

For regional studies of climatic variability with interest in spatial and/or spatio-temporal patterns the climatic series to be used are time series of bi-dimensional almost horizontal fields (surface or constant pressure levels). These fields can be specified by a number of irregularly spaced time series in meteorological stations or, preferably, in regular grids. The gridded values are the result of the averaging of synoptic meteorological analysis based in the observations in the region and performed by methods that use interpolation with physical consistency (e.g., Kalnay et al., 1996; Kistler et al., 2001). The term reanalysis refers to *ad posteriori* analyses using constant procedures over a large period of time, to avoid inhomogeneities that could otherwise be introduced in the series by the modification of the analysis methods over time.

Principal Component Analysis (PCA) is commonly used in understanding the principal modes of climatic variability of an atmospheric variable. PCA is based on a compression method that reduces the variability to some number of modes explaining a considerable part of the field variance. A detailed analysis of the mathematics of the Empirical Orthogonal Functions (EOFs) approach to both scalar and vector fields can be found in Peixoto and Oort (1992). In von Storch and Zwiers (1999) is presented also a detailed description of the technique with applications in climatology. The method is applied on this study, both to a small country or region (e.g. Portugal, in case of winter seasonal precipitation) and to an extended area like the atmospheric surface pressure over the North Atlantic, allowing the knowledge of the most important variability modes, namely the well known leading mode of climate variability over the North Hemisphere, the North Atlantic Oscillation (NAO).

The Singular Spectral Analysis (SSA) is a recent tool used in time series analysis. Contrary to classical spectral analysis or maximum entropy methods (MEM), where the basis functions are prescribed sines and cosines, SSA produces data-adaptive filters that are able to isolate oscillations spells, which makes the method more flexible and better suited for the analysis of non-linear, anharmonic oscillations (Vautard et al., 1992). Furthermore, SSA allows the decomposition of the temporal series associated to each mode in trends and periodicity component series. These series are easier to analyse due to the obvious noise reduction and permits the grouping of selected temporal reconstructed components of the detected modes.

However, PCA just allows the evaluation of the principal spatial modes and how the respective spatial patterns evolve in time. The quasi-meridional NAO behaviour has been reported in several studies (Hurrell, 1996; Hurrell & van Loon, 1997) where analyses are performed by methods just based on PCA. However some authors used or are using, depending on the season, different locations mainly for the southern pole of the NAO index: Azores (Rogers, 1984), Lisbon (Antunes et al., 2006) and Gibraltar (Jones et al., 1997). These different locations are justified by the non-stationary seasonal behaviour of the poles of the oscillation that is evident when seasonal PCA is performed seasonally (e.g. Hurrell et al., 2003).

The analysis of space-temporal variability, that is, the spatio-temporal analysis of modes evolving in space and time may provide more information about the underlying physical system (Vautard, 1995). In fact, a spatio-temporal analysis of the North Atlantic mean sea level pressure field reveals that the first mode of variability only behaves like what was supposed to be the NAO pattern, in extreme high and low NAO index phases. This space-time analysis shows that the principal mode of oscillation is not always quasi-meridional, as it is frequently assumed, but has an oscillation pattern that changes assuming different orientations with time (Antunes et al., 2010).

The application of PCA to the annual or seasonal precipitation amount in Portugal reveals very simple variability patterns (Serrano et al., 2003). In winter the consideration of just three Empirical Orthogonal Functions (EOFs) explains 94% of the total variability and the first mode, associated with precipitation anomalies of the same sign over the whole region, explains 85% of the total variance.

The analysis performed by SSA is able to detect periodic characteristics in the time variability of the precipitation amount (Antunes et al., 2000) but these signals are not found to be statistically significant. However, when a space-temporal method is applied, the space time periodic signals became statistically significant, revealing again that the Multichannel Singular Spectral Analysis (MSSA) is an efficient tool in the detection of non-purely random characteristics in series with a weak signal/noise relation.

All the signal components, detected by SSA or MSSA, must be tested for their significance to determine if they correspond to real oscillations. Monte Carlo analysis is a recognized method to estimate the confidence intervals of determined significance levels. The tests formulated by Allen and Robertson (1996) assume that data have been generated, according to the null hypothesis, by first-order autoregressive (AR(1)) independent processes (red noise).

Since a large fraction of the surface atmospheric pressure field variance in the North Atlantic can be explained by the NAO (Hurrel, 1996), and considering that the western Portugal is located near the Atlantic Ocean, it is not surprising that very significant correlations can be found between pressure and precipitation amount climatic series.

Recently, several studies tried to detect the Atlantic variability modes (spatio or temporal) at several time scales, and their relations with precipitation variability for the whole Iberian Peninsula, mostly in winter, when the great part of precipitation occurs (e.g. Zorita et al., 1992; Esteban-Parra et al., 1998; Rodriguez-Puebla et al., 1998; Ulbrich et al., 1999; Goodess & Jones, 2002; Trigo et al., 2004).

The analyses using selected components extracted from the spatio-temporal variability reinforce the importance of detection of these noise reduced signals with similar periodic characteristics. The cross correlation functions reveal that there is a potential predictability of winter precipitation in Portugal.

The spatial evolution of the principal variability modes detected by MSSA also allows the knowledge improvement of the behaviour of both variables for the same time steps.

2. Data

The analysis of mean sea level pressure fields was performed using data from the National Centers for Environmental Prediction/National Center for Atmospheric Research

(NCEP/NCAR) reanalysis project from 1949 to 2000 (Kalnay et al., 1996). Data are available in a latitude–longitude grid (2.5° × 2.5°) and refer to monthly means that are subsequently processed resulting in annual and seasonal means. The analysed area of the North Atlantic is bounded by latitudes 20 and 80°N and longitudes 90 and 0°W. Data obtained from reanalysis tend to be spatial-scale dependent causing the violation of the assumption of independence required for the statistical analysis. Thus, to avoid this dependence in adjacent series and the consequent temporal autocorrelation, the first ten principal components obtained by PCA are used instead of all channels of the field grid points. The former are orthogonal, that is, non-correlated for zero time lags.

The analysis of multivariate precipitation time series was performed using 11 stations located in mainland Portugal from 1949 to 2000. The data represent the highest quality-controlled series recorded in Portugal. These data have no missing values and have been subjected to a previous homogeneity analysis. More details about localization, station metadata and a climatologic summary can be found in Antunes (2006). Before processing, the data were normalized, that is, centred around the mean and standardized by the standard deviation.

For both variables, winter refers to the months from December to February. The other seasons are defined similarly and the annual means are calculated from January to December.

3. Methods

3.1 Multichannel singular spectral analysis

PCA is frequently used to compress variability into a reduced number of modes that explain a considerable part of the field variance. However, this method just allows the evaluation of the spatial mode and shows how the respective spatial pattern evolves in time, that is, the EOFs and the associated T-PCs.

The space-temporal variability can be accomplished through the use of multichannel singular spectral analysis (MSSA) which is an extension of SSA, used in time series analysis and described in Vautard et al. (1992).

Considering a spatio-temporal field X_{li}, l being the spatial index ($l \leq l \leq L$), i the discrete temporal index ($1 \leq i \leq N$) and j the temporal lag, MSSA can be formulated as

$$X_{l,i+j} = \sum_{k=1}^{L \times M} a_i^k E_{lj}^k \qquad 1 \leq l \leq L, \quad 1 \leq j \leq M$$

where the coefficients a_i^k are the spatio-temporal principal components (ST-PCs) and E^k are the eigenvectors of the cross-covariance matrix called spatio-temporal empirical orthogonal functions (ST-EOFs).

The MSSA method computes the lagged cross covariances between the channels, and a multichannel trajectory matrix is created. This is done by first generating each channel (i.e. either the time series of each grid point or the time series of each T-PC) with M lagged copies of itself, M being the number of temporal lags, and then forming the full augmented trajectory matrix (Allen & Robertson, 1996).

This lagged cross covariance matrix T_X of dimension (LxM)x(LxM) is formed by blocks that contain the lagged cross covariances of pairs of channels at lags 0 to M-1:

$$T_X = \begin{pmatrix} T_{11} & T_{12} & \cdots & \cdots & & T_{1L} \\ T_{21} & T_{22} & \cdots & & \cdots & \cdots \\ \cdots & \cdots & \cdots & \cdots & & \cdots \\ \cdots & \cdots & \cdots & T_{ll'} & & \cdots \\ \cdots & \cdots & \cdots & & \cdots & T_{L-1L} \\ T_{L1} & \cdots & \cdots & T_{LL-1} & & T_{LL} \end{pmatrix}$$

where $T_{ll'}$ of dimension (MxM) is the lag covariance of X between channels l and l'.

Singular value decomposition (SVD) was the method applied to this matrix to yield two orthogonal bases, the right and the left singular vectors (Robertson, 1996). The first of these, the eigenvectors of the matrix, are a sequence of spatial modes (ST-EOFs) and the others, the associated principal components (ST-PCs), represent the way how these patterns evolve in time. The orthogonality of the covariance matrix both in time (zero cross covariance of two different ST-PCs at lag 0) and in space (orthogonality of the ST-EOFs) implies that the diagonal of the matrix corresponds to the associated explained variances. These variances are of decreasing importance as the order k increases.

More details of MSSA can be found in Plaut and Vautard (1994).

Both SSA and PCA are particular cases of the MSSA. SSA can be derived considering just one series in analysis:

$$x_{i+j} = \sum_{k=1}^{M} a_i^k E_j^k = X_{ji} \qquad 1 \le j \le M$$

and PCA can be derived from MSSA considering the null time lag:

$$X_{li} = \sum_{k=1}^{L} a_i^k E_l^k \qquad 1 \le l \le L$$

MSSA, being a spatio-temporal variability analysis and allowing the evolution of the spatial patterns in time, improves the knowledge of modes evolving simultaneously in space and time. This analysis solves an eigenvalue matrix problem to be solved of an added embedding dimension of (number of series * window length) (Plaut & Vautard, 1994).

As in other methods of spectral analysis, the window length M is an essential issue in MSSA. Its selection involves a compromise: N being the number of terms of the time series, $N - M + 1$ needs to be large enough to allow adequate signal/noise enhancement, but small enough to guarantee the statistical robustness. In this and previous studies (Antunes et al., 2006, Antunes et al., 2010), the results of the experiments performed by Plaut and Vautard (1994) have been considered: the method does not distinguish between different oscillations of period longer than M, and a window length M typically allows the distinction of oscillations of periods in the range $(M/5, M)$.

3.2 Statistical significance testing

Contrary to classical or maximum entropy spectral analysis, where the basis functions are prescribed sines and cosines, SSA produces data-adaptive filters that are able to isolate oscillations spells, which makes the method more flexible and better suited for the analysis of non-linear, anharmonic oscillations (Vautard et al., 1992). Similar to SSA, if the variance of the series is dominated by an oscillation, MSSA generates a pair of EOFs nearly periodic, with a similar period, and in quadrature (Plaut & Vautard, 1994).

However, these pairs can also be randomly generated realizations of processes which do not have an oscillatory nature and consequently, once detected, must be tested to determine if they correspond to real oscillations.

The null-hypothesis, as formulated by Allen and Robertson (1996), is that data have been generated by L first-order autoregressive (AR(1)) independent processes (red noise), with L being the number of channels.

The null-hypothesis can be rejected if the spectrum of the eigenvalues associated with the modes detected by MSSA is higher than that expected in data generated by red noise processes. The confidence intervals for a given significance level are estimated using Monte Carlo simulations. A large ensemble of normally distributed (gaussian) surrogate noise time series is generated with the null hypothesis characteristics, in the case autoregressive first order processes, with the same length, variance and temporal lag-one autocorrelation as the series that form the centred input channels.

Allen and Robertson (1996) suggest two different significance tests based on the same null-hypothesis: the first using the ST-PCs and ST-EOFs of the data, and the second using the ST-PCs and ST-EOFs of the null hypothesis. In the first test the lag covariance matrix is computed from the data whereas, in the second, the lag covariance matrix is computed from data generated by Monte Carlo methods. The subsequent procedure of projecting data and noise surrogates onto the vector basis is similar in both the tests. The second test is more robust as the former implicitly assumes the existence of a signal before any signal has been identified (Allen and Smith, 1996). Besides that, the second test does not present the artificial variance compression problem that boosts the significance of the first ST-PCs in the detriment of higher order modes.

Tests are based on the WMO Technical Note N° 79 (Mitchell, 1966): a spectral peak is considered significant at the 0.05 level if it goes above the upper limit of the 90% confidence interval with limits 5 and 95% (unilateral or one-tailed test).

4. Pressure variability in the North Atlantic

A Principal Component Analysis (PCA) applied to the mean sea level pressure field of the North Atlantic reveals the North Atlantic Oscillation (NAO) pattern as the principal mode of annual and seasonal variabilities. Before the application of PCA the time series were normalized, that is, centred and standardised to avoid overweighting the locations with larger variance; linear trends for each point of the field were also removed.

This oscillation mode, presented in Figure 1 for all seasons, means that pressure above the mean at mid-latitudes tends to occur simultaneously with pressure below the mean at higher latitudes (positive NAO phase) and vice-versa, pressure below the mean at mid-latitudes tends to occur with pressure above the mean at higher latitudes (negative NAO phase).

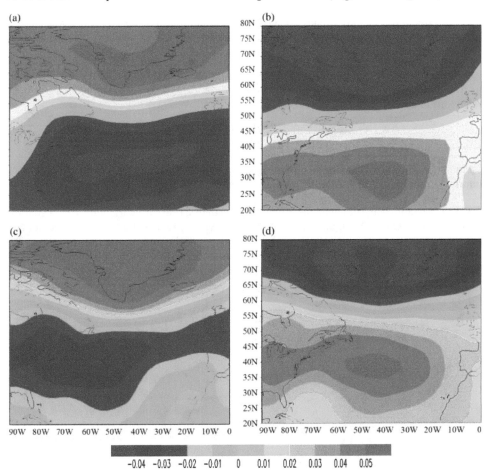

Fig. 1. Representation of the first EOFs of North Atlantic pressure variability in (a) winter, (b) spring, (c) summer and (d) autumn. The units are arbitrary.

The NAO is the principal mode of variability in all seasons but the patterns, namely the locations of the centres of action, differ from season to season. The variance explained by each first seasonal mode also changes depending on the season: it is in winter and in spring that these principal modes explain the most variance of the field variability (43 and 41%, respectively); in summer and autumn these modes just explain 32 and 30%, respectively, of the total pressure variance.

The analysis of the time variability of the temporal principal components (T-PCs) associated with each principal seasonal empirical orthogonal functions (EOFs) was performed using Singular Spectral Analysis (SSA). The same analysis was applied to the T-PC associated with the first annual pressure pattern. The method provides the decomposition of each of these series in reconstructed components, allowing to isolate the part of the signal involved with an oscillation. Table 1 presents the periodicities, estimated by maximum entropy methods, of the first two paired components of each T-PC. Results reveal different periodicities in the first annual and seasonal modes, which are not statistically significant when tested against the red noise null hypothesis at the 0.05 significance level.

Comparison between the periodicities shows that winter is the season with most influence on the annual first mode of the pressure field variability. This season reveals a periodic behaviour of about 9 years.

	Year	Winter	Spring	Summer	Autumn
Periodicity (years)	13.5/15.0	8.8/9.1	4.3/4.4	5.6/6.9	2.1/2.2

Table 1. Periodicities of the first two paired components extracted by SSA from the first annual and seasonal North Atlantic pressure variability modes.

The winter spatio-temporal analysis of the North Atlantic pressure variability performed by the use of multichannel singular spectral analysis (MSSA) reveals that the first two spatio-temporal principal components (ST-PCs) are nearly periodic and in quadrature (Figure 2), satisfying the conditions to represent an oscillation, according to the Plaut and Vautard (1994) criteria.

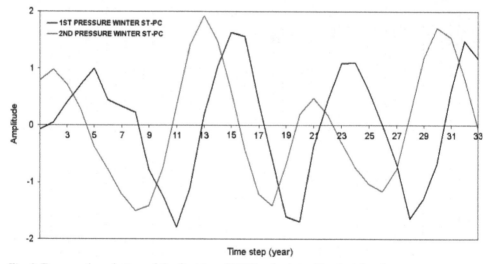

Fig. 2. Temporal evolution of the first two ST-PCs of winter North Atlantic pressure.

They also satisfy a third more restricted condition proposed by the same authors which establishes the existence of the oscillatory pair: in order to guarantee that, at least, one cycle of the oscillation is coherent, it is required to have a periodic behaviour in the ST-PCs cross correlation function with correlation absolute values higher than 0.5 for the two successive extremes on each side of the lag zero (Figure 3). According to Bartlett, the cross correlation confidence intervals corresponding to the 0.05 significance level, estimated from the normal distribution with a standard deviation of $\pm 1*\mathrm{sqrt}(N - |\tau|)$, where τ represents the temporal lag, is also presented in this figure (von Storch & Zwiers, 1999).

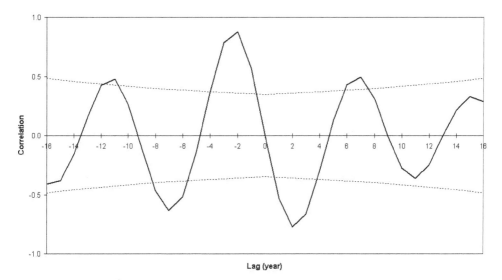

Fig. 3. Lagged cross correlation function between the first and second ST-PCs of winter North Atlantic pressure and confidence intervals of the 0.05 significance level.

Following the methodology presented in Section 3.2, the statistical significance of the oscillations is tested against the null-hypothesis of red noise, at the 0.05 significance level, using Monte Carlo methods. The results, using a base derived from data and also a base derived from the null-hypothesis, are presented in Figure 4(a) and (b), respectively. In the first case, three pairs of significant eigenvalues are detected: the longest significant components show a periodicity of about 9 years, followed by 4.5 and 2.7 years periodicity modes. However, in a more conservative significance test that uses a base derived from AR(1) processes, the analysis just shows the 2.7 years periodicity as significant. Other authors analysing the Northern Hemisphere sea level pressure have already detected this 2/3 years shorter periodicity, but the mechanism behind this peak remains uncertain (Stephenson et al., 2000). Besides that, in the second significance test, which is more stringent than the first because the null-hypothesis is more difficult to reject, the eigenvalue associated with a 9 years periodicity has a spectral density almost reaching the 0.05 significance upper level.

The 9 years spatio-temporal mode is the mode which explains the largest fraction of the pressure winter field variance (15%). Power with this periodicity has already been detected (although not statistically significant) in the T-PCs derived from a simple PCA of the same field.

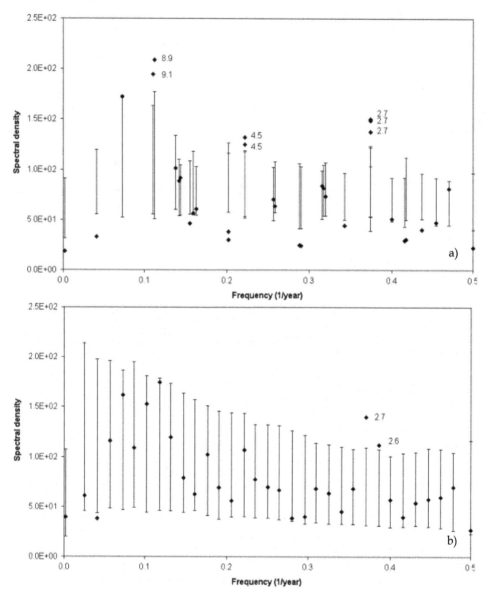

Fig. 4. Monte Carlo significance tests for winter pressure oscillations using a window length $M = 25$. The vertical bars indicate the 0.05 significance level (one-tailed test) for the red noise null-hypothesis. Periodicities of significant eigenvalues for this hypothesis are also shown. (a) Diamonds show the data eigenvalues projected onto the data-adaptive basis. (b) Diamonds show the data eigenvalues projected onto the null-hypothesis basis.

This winter mode of spatio-temporal variability of about 9 years periodicity in North Atlantic is presented in Figure 5, where the time origin is arbitrary.

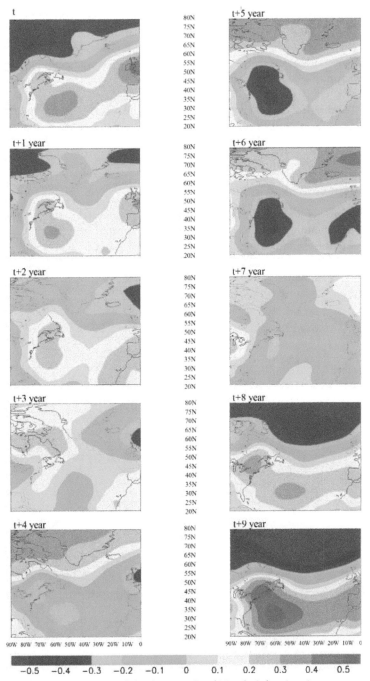

Fig. 5. Spatio-temporal evolution of the first mode of North Atlantic winter pressure variability for 1 year lags. The units are arbitrary.

The mode of oscillation has a quasi-meridional pattern during the high and low NAO index phases, with positive anomalies in the central North Atlantic and negative anomalies at high latitudes. The transition from the positive to the negative phase shows the weakening of pressure at lower latitudes and the strengthening at higher latitudes till half of the period. After that, the opposite evolution occurs, with pressure strengthening at lower latitudes and weakening at higher latitudes. During the cycle, for some lags (e.g. temporal lags of 3 and 7 years) corresponding to transition phases, the patterns are quite different from what is known to be the NAO pattern. In the first case, the principal dipole reveals a W/E orientation at higher latitudes; in the second transition year, there is a diffuse pattern with no clear defined centres of action.

5. Precipitation variability in Portugal

The longest records of annual and seasonal precipitation time series of Portuguese stations, analysed by maximum entropy methods (MEM) do not reveal any characteristics significantly different from white-noise processes (Antunes & Oliveira Pires, 1998). Figure 6 presents, as an example, the variance spectrum for the Lisbon annual precipitation time series (1871/1993) estimated by MEM. In same figure are also presented the 0.1 confidence intervals for the white noise null hypothesis.

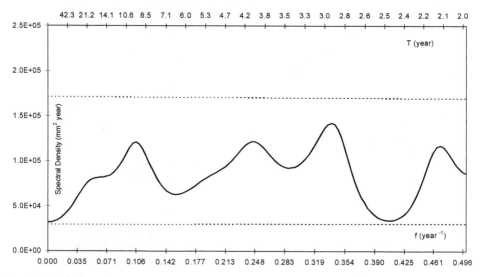

Fig. 6. Lisbon annual precipitation variance spectrum and the 90% confidence intervals for the null hypothesis of white noise.

The singular spectral analysis (SSA) applied to the same series is also unable to detect significant characteristics in precipitation variability (Antunes et al., 2000). Figure 7 presents the singular values estimated by this method, the associated error bars and the 90% confidence intervals estimated using the Monte Carlo methods for the white noise null hypothesis.

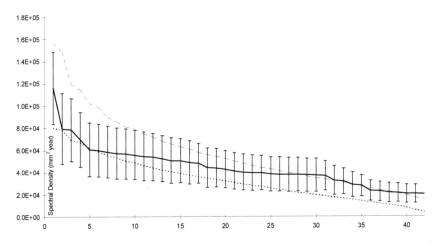

Fig. 7. Eigenvalue spectrum of annual precipitation amount in Lisbon, error bars and the 0.1 significance level for the white noise null hypothesis.

The spatial variability analysis of annual and seasonal precipitation in Portugal using 11 meteorological stations of more than 50 years, performed by PCA, reveal that the first variability mode, that shows precipitation varying in same way in all country, explains a large fraction of the variance (75% in annual precipitation and 85% in winter which is the rainier season). Figure 8 presents the first three Empirical Orthogonal Functions (EOFs) for winter. The second EOF, explaining 6% of variance shows an opposite behaviour between north and southern regions and the third EOF accounting for 3% of total variability shows the opposite relation between littoral and inland, more obvious in northern areas. The second EOF shows that precipitation above the mean in northern areas of the country tends to occur with precipitation below the mean in southern areas and vice-versa; the third EOF reveals the opposite, coastland precipitation above the mean tending to occur with inland precipitation below the mean and vice-versa.

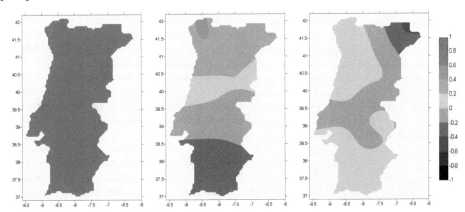

Fig. 8. Principal empirical orthogonal functions of winter precipitation in Portugal.

The analysis of the temporal principal components (T-PCs) associated to each winter spatial mode of precipitation variability does not reveal significant signals in relation to the red noise null hypothesis or even considering the white noise null hypothesis.

Despite not being statistically significant is interesting the results of the singular spectral analysis (SSA) applied to the first T-PC associated to the first winter EOF mode. The method detects in this T-PC two oscillatory pairs that explain 39% of the total variance of the component. The spectral analysis of the reconstructed components (RCs) that form these pairs reveal obvious spectral characteristics around the same frequencies. The principal peaks occur in the 8 years band and the second pair presents shorter periodicity characteristics of about 2.8 years (Figure 9).

Similar oscillatory components of about 9 and 2.7 years periods were detected in the annual precipitation variability (Antunes, 2007). The application of Monte Carlo methods reveals that these signals are not significant, just like in the winter variability. In the other seasons the precipitation variability is characterized by shorter periodic fluctuations, being the signals with longer period comprised between 3.4 and 4.6 years. These results show that the principal mode of annual variability is dominated by the principal precipitation variability mode in winter.

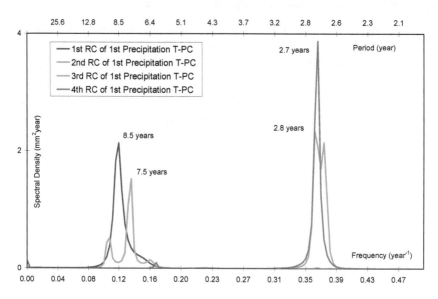

Fig. 9. Variance spectra of the reconstructed components (RCs) extracted by SSA from the first temporal principal component (T-PC) of winter precipitation in Portugal. Main periodicities of each RC are included.

The analysis of winter precipitation temporal variability performed by the use of patterns that evolve spatially allows the detection of periodic signals similar to those already obtained by the methods previously presented. Moreover the consistency of the results, the method identifies these signals as significant, suggesting a more realistic analysis process by allowing the change of the EOFs configuration with time.

The application of the Multichannel singular spectral analysis (MSSA) to the winter precipitation variability reveals the existence of spatio-temporal principal components (ST--PCs) that, due to their similar periodicity, can form oscillatory pairs. Figure 10 shows a pair of these ST-PCs that is nearly periodic and in quadrature, satisfying the conditions to represent an oscillation according to the Plaut and Vautard (1994) criteria.

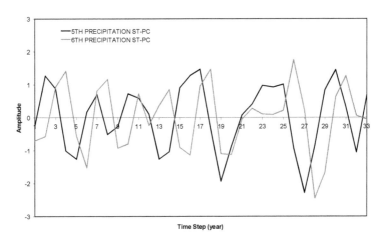

Fig. 10. Temporal evolution of the 5 and 6th winter precipitation ST-PCs in Portugal.

These components also satisfy a third more stringent condition proposed by the same authors which establishes the existence of the oscillatory pair: in order to guarantee that at least one cycle of the oscillation is coherent, a periodic behaviour in the ST-PCs cross correlation function is required, with the absolute values of correlation higher than 0.5 for the two successive extremes on each side of lag zero (Figure 11).

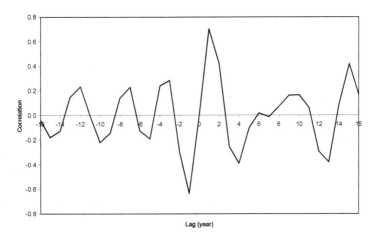

Fig. 11. Lagged cross correlation function between the 5 and 6th winter precipitation ST--PCs in Portugal.

In Figures 12 and 13 are presented the corresponding results of another pair of components with shorter periodicity that satisfies the same criteria.

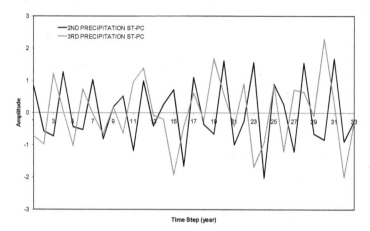

Fig. 12. Temporal evolution of the 2nd and 3rd winter precipitation ST-PCs in Portugal.

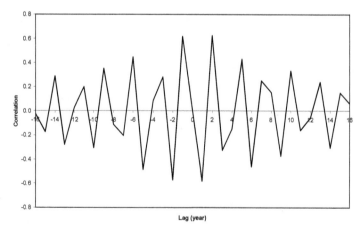

Fig. 13. Lagged cross correlation function between the 2nd and 3rd winter precipitation ST--PCs in Portugal.

Following the methodology presented in Section 3.2, the statistical significance of these oscillations was tested by Monte Carlo methods in relation to the null-hypothesis of red noise at the 0.05 significance level (one tailed test). Results using a base derived from data and a base derived from the null-hypothesis are presented in Figure 14(a) and 14(b), respectively. In the first case, two pairs of significant eigenvalues are detected. The oscillation that presents the higher spectral density reveals a period of about 2.7 years. The second oscillatory mode, consisting of ST-PCs 5 and 6, reveal a period of about 7 years. The eigenvalue associated to a period of about 9 years (ST-PC 4) could not be paired with the previous ones of about 7 years periodicity. The second test confirms the occurrence of a significant signal with a period of about 2.7 years; a signal of 8.4 years periodicity is however near the superior limit of the significance level.

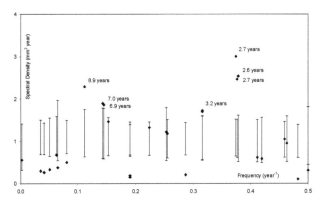

Fig. 14. (a)- Monte Carlo significance tests for winter precipitation oscillations. Diamonds show the data eigenvalues projected onto the data-adaptive basis. The vertical bars indicate the 0.05 significance level (one-tailed test) for the red-noise null-hypothesis. Periodicities of significant eigenvalues for this hypothesis are also shown.

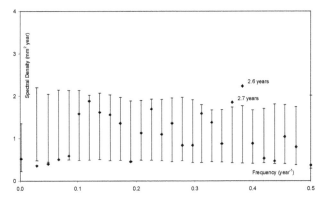

Fig. 14. (b)- Monte Carlo significance tests for winter precipitation oscillations. Diamonds show the data eigenvalues projected onto the null-hypothesis basis. The vertical bars indicate the 0.05 significance level (one-tailed test) for the red-noise null-hypothesis. Periodicities of significant eigenvalues for this hypothesis are also shown.

Although the second, more conservative significance test, just reveals the oscillation with shorter period as significant, the detected periodicity of about 7/9 years must be taken into account. In fact, the same analysis applied to the annual precipitation in Portugal also detects a periodic signal about the 9 years in its variability (Antunes et al, 2006.). In this case it corresponds to the first variability mode, which presents the higher spectral density, explaining 21% of the total annual spatio-temporal variance.

Figure 15 shows the space-time evolution of the winter ST-EOF associated to the ST-PC presenting the 9 years periodicity, for one-year steps during 16 years.

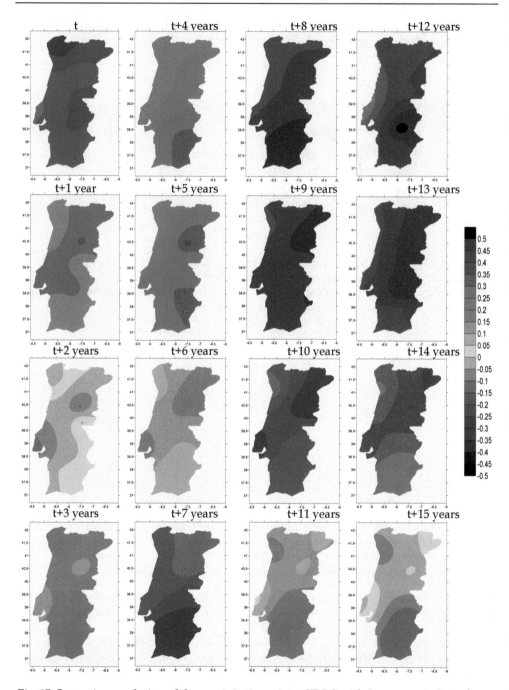

Fig. 15. Space-time evolution of the precipitation winter ST-PC with 9 years periodicity for one year-lags during 16 years. The units are arbitrary.

The most frequent pattern is, as in the PCA, the occurrence of precipitation anomalies of the same sign in the whole region. The spatial evolution of this mode is towards the occurrence of precipitation above the mean during a period of about 4 years following a period of the same duration with precipitation below the mean plus, when existing, a transition period of 1 year to complete the 9 years cycle. A complete period in the figure is identified, for example, by years (t + 7) to (t + 10) with precipitation below the mean, following precipitation above the mean from year (t + 11) to (t + 14) and a transition year represented by the (t + 15) time step. In transition years, the patterns present a same sign zone oriented NW/SE or SW/NE and the opposite sign in the rest of the region.

6. Relations between pressure and precipitation winter variability in Portugal

The possibility of reconstructing the original field for a period equal to the series under consideration based on their spatio-temporal characteristics has advantages over the use of the ST-PCs that are shorter series, with length equal to the time series subtracted by the width of the window. In order to analyse the relations between the pressure in the Atlantic and the precipitation in Portugal, the reconstruction of these fields was performed using a successively larger number of space-time components that successively explain more variance of the fields. These components are reconstructed in the case of pressure to a grid point (40° N, 7.5° W) in Portugal near Lisbon and in the case of precipitation to the point located in Lisbon. The reconstruction performed using 20 components explains 85.5% of the total variance of the winter pressure field in the North Atlantic and the reconstruction of precipitation with the same number of components explains 93.8% of the winter variability.

The use of a great number of reconstructed components (RCs) allows the higher zero lag correlations. However, whatever the number of components in the analysis, these correlations are always significant, ranging between -0.6 and -0.8. As can be seen in Fig 16 there are also significant cross-correlations, with the same sign, for lags of 8 years, which means that there is a periodic correlation of almost sinusoidal type about 8 years between the principal variability modes of pressure and precipitation. Related to the same modes other significant correlations of opposite sign are detected to lags of ±4 and ±12 years. Other significant correlations that occur for other lags result from the sum of components with different frequencies.

The analysis of the same figure reveals that for lags different than the null, it verifies higher correlations (in module) when is used a smaller number of components. However, the smaller the number of components considered, the lower the variance explained of the field.

As a result from the lagged cross correlation functions analysis it can be concluded that the winter pressure field can be used as a good predictor of the precipitation field in the same season in Portugal, since the field itself is capable to be forecasted.

It is also interesting the joint analysis of the pressure and precipitation oscillatory spatio-temporal modes that present the higher spectral density and associated to the 9 years periodicity (fig. 4.a) and 2.7 years (fig. 14.a), respectively. The cross correlation function between these modes reveals significant and positive values for the null lag (Fig.17). The positive signal is important in interpreting the relations between the evolution modes presented in Figure 18, for simultaneous years.

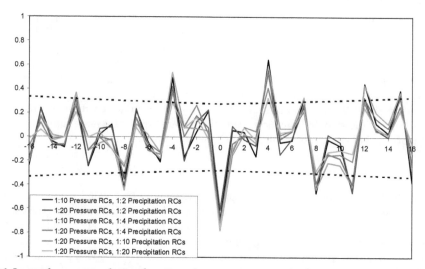

Fig. 16. Lagged cross correlation functions between reconstructed component groups of North Atlantic winter pressure and of Portugal winter precipitation. Confidence intervals of the 0.05 significance level are also presented. Positive lag values correspond to delays of the precipitation reconstructed component groups.

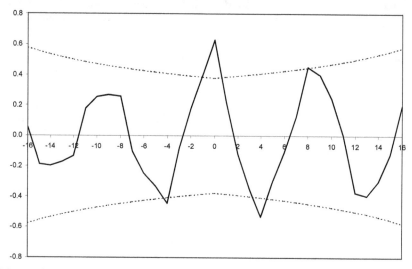

Fig. 17. Lagged cross correlation function between the first winter pressure and precipitation ST-PCs and confidence intervals of the 0.05 significance level.

In the major part of the considered time steps it appears, for the same year, the occurrence of the NAO in positive phase and precipitation below the mean value in Portugal and vice-versa, that is, the occurrence of the NAO negative phase and precipitation above the mean. These relations can be observed in the same figure taking into account the significant correlation obtained for 4 years lags and considering the negative value of this correlation.

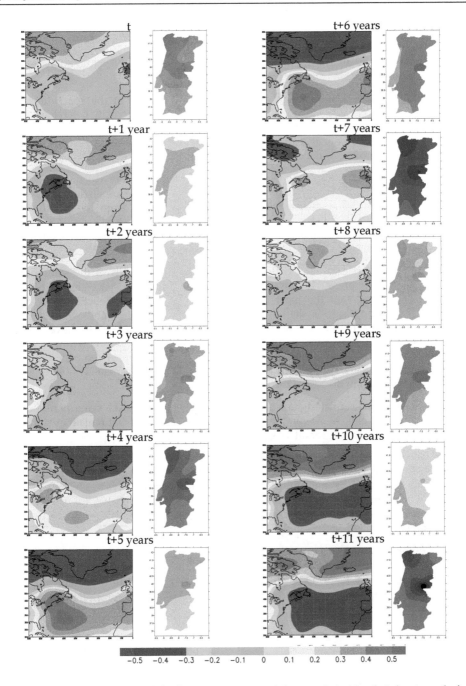

Fig. 18. Space-time evolution of the first pressure variability mode in North Atlantic and of the first precipitation variability mode in Portugal, in winter during 12 years.

7. Conclusions

The North Atlantic mean sea level pressure and the Portuguese precipitation winter variability fields were analysed using several methods including the spatio analysis (PCA), the temporal analysis (SSA) and the spatio-temporal analysis (MSSA).

The application of the multichannel singular spectral analysis in the space–time study allowed the detection of significant modes of variability with periodicity behaviour, both in pressure and in precipitation. These periodic behaviours were previously detected in the temporal principal components extracted from the principal component analysis but, in the time domain, were found as not statistically significant.

In the North Atlantic pressure field variability the detected signals revealed a periodicity of about 9 years. The representation of this mode shows that the oscillation is not quasi-meridional but has different orientations, rotating in a cycle, from the positive North Atlantic oscillation (NAO) phase through the negative NAO phase and again to the positive phase. This principal variability mode only behaves like the traditional NAO pattern in extreme high and low NAO index phases.

The precipitation variability in Portugal reveals periodicity signals of the same order of those detected in the pressure analysis. The representation of the space–time evolution of one of modes reveal that the most frequent pattern is the occurrence of precipitation anomalies of the same sign in the whole region, that is, precipitation above or below the mean. The spatial evolution is towards the occurrence of precipitation above the mean during a period of about 4 years following a period of the same duration with precipitation below the mean plus, when existing, a transition period of 1 year to complete the 9 years cycle. In transition years, the patterns present a same sign zone oriented NW/SE or SW/NE and the opposite sign in the rest of the region.

It were also detected similar signals of shorter periodicity, about 2.7 years periodicity, on variability of both fields. However, these are precisely the signals that remain as statistically significant when a more conservative significance test is applied.

The relations between the North Atlantic mean sea level pressure and the Portuguese precipitation winter variability fields were analysed using groups of reconstructed components estimated by the multichannel singular spectral analysis. The results reveal significant correlations for null lags, even when just a few components are used, meaning that the pressure field can be a good predictor of the precipitation field, since the field itself is predictable and may be forecasted.

The space-time simultaneous evolution of the pressure and precipitation modes with higher spectral density reinforce the importance of this relationship revealing for the same time step the occurrence of the NAO positive phase and precipitation below average in Portugal and vice-versa, the occurrence of the NAO negative phase and precipitation above the mean.

8. References

Allen, M. & Robertson, A. (1996). Distinguishing modulated oscillations from coloured noise in multivariate datasets. *Climate Dynamics*, 12, 775-784.

Allen, M. & Smith, L. (1996). Monte Carlo SSA: detecting irregular oscillations in the presence of colored noise. *Journal of Climate*, 9, 3373-3404.

Antunes, S., Oliveira Pires, H. & Rocha, A. (2010). Spatio-Temporal Patterns of Pressure over the North Atlantic. *International Journal of Climatology*, 30, 2257-2263.

Antunes, S. (2007). *Variabilidade climática no Atlântico e suas relações com o clima de Portugal.* Ph.D. thesis, University of Aveiro (in Portuguese).

Antunes, S., Oliveira Pires, H. & Rocha, A. (2006). Detecting spatio-temporal precipitation variability in Portugal using multichannel singular spectral analysis. *International Journal of Climatology*, 26, 2199–2212.

Antunes, S., Oliveira Pires, H. & Rocha, A. (2000). Singular Spectral Analysis (SSA) applied to Portuguese temperature and precipitation amount series, *Proceedings on 15th Conference on Probability and Statistics in the Atmospheric Sciences*, American Meteorological Society, Asheville, North Carolina.

Antunes, S. & Oliveira Pires, H. (1998). Contribution to the characterization of the interannual climatic variability in Portugal, *Proceedings of 1st Symposium on Meteorology and Geophysics*, Associação Portuguesa de Meteorologia e Geofísica, ISBN 972-8445-12-1, Lagos, Portugal (in Portuguese).

Duplessy, J. & Morel, P. (1990). *Gros temps sur la planète*, Édition Odile Jacob, ISBN 2.02.016520.1, Paris.

Esteban-Parra, M., Rodrigo, F., Castro-Diez, Y. (1998). Spatial and temporal patterns of precipitation in Spain for the period 1880-1992. *International Journal of Climatology*, 18, 1557-1574.

Goodess, C. & Jones, P. (2002). Links between circulation and changes in the characteristics of Iberian rainfall. *International Journal of Climatology*, 22, 1593-1615.

Hurrell, J. (1996). Influence of variations in extratropical wintertime teleconnections on Northern Hemisphere temperatures. *Geophysical Research Letters*, 23, 665-668.

Hurrell, J., Kushnir, Y., Ottersen, G. & Visbeck, M. (2003). *The North Atlantic Oscillation: Climate significance and environmental impact.* Geophysical Monograph Series, 134. American Geophysical Union: Washington.

Hurrell, J. & van Loon, H. (1997). Decadal variations in climate associated with the North Atlantic Oscillation. *Climate Change*, 36, 301-326.

Jones, P., Briffa, K., Osborn, T., Lough, J., van Ommen, T., Vinther, B., Luterbacher, J., Wahl, E., Zwiers, F., Mann, M., Schmidt, G., Ammann, C., Buckley, B., Cobb, K., Esper, J., Goosse, H., Graham, N., Jansen, E., Kiefer, T., Kull, C., Küttel, M., Mosley-Thompson, E., Overpeck, J., Riedwyl, N., Schulz, M., Tudhope, A., Villalba, R., Wanner, H., Wolff, E. & Xoplaki, E. (2009). High-resolution palaeoclimatology of the last millennium: A review of current status and future prospects. *The Holocene*, 19, 3–49.

Jones, P., Jonsson, T. & Wheeler, D. (1997). Extension to the North Atlantic oscillation using early instrumental pressure observations from Gibraltar and South-West Iceland. *International Journal of Climatology*, 17, 1433–1450.

Kalnay, E., Kanamitsu, M., Kistler, R., Collins, W., Deaven, D., Gandin, L., Iredell, M., Saha, S., White, G., Woollen, J., Zhu, Y., Leetmaa, A., Reynolds, R., Chelliah, M., Ebisuzaki, W., Higgins, W., Janowiak, J., Mo, KC., Ropelewski, C., Wang, J., Jenne, R., Joseph, D. (1996). The NCEP/NCAR 40-year reanalysis project. *Bulletin of the American Meteorological Society*, 77, 437–471.

Kistler, R., Kalnay, E., Collins, W., Saha, S., White, G., Woollen, J., Chelliah, M., Ebisuzaki, W., Kanamitsu, M., Kousky, V., van den Dool, H., Jenne, R. & Fiorino, M. (2001). The NCEP-NCAR 50-year reanalysis: monthly means CD-ROM and documentation. *Bulletin of the American Meteorological Society*, 82, 247-268.

Lee, T., Zwiers, F. & Tsao, M. (2008). Evaluation of proxy-based millennial reconstruction methods. *Climate Dynamics*, 31, 263–281.

Ljungqvist, F. (2010). A Regional Approach to the Medieval Warm Period and the Little Ice Age. *Climate Change and Variability*, Suzanne Simard (Ed.), ISBN: 978-953-307-144-2, InTech.

Mann, M. & Jones, P. (2003). Global surface temperatures over the past two millennia. *Geophysical Research Letters*, 30, 1820-1823.

Mitchell, J. (1966). *Climatic change*, WMO Technical note n° 79.

Osborn, T. & Briffa, K. (2006). The spatial extent of 20th-century warmth in the context of the past 1200 years. *Science*, 311, 841–844.

Peixoto, J. & Oort, A. (1992). *Physics of Climate*, American Institute of Physics, ISBN 0-88318-711-6, New York.

Plaut, G. & Vautard, R. (1994). Spells of low-frequency oscillations and weather regimes in the northern hemisphere. *Journal of the Atmospheric Sciences*, 51-2, 210–236.

Robertson, A. (1996). Interdecadal variability over the North Pacific in a multi-century climate simulation. *Climate Dynamics*, 12, 227–241.

Rodriguez-Puebla, C., Encinas, A., Nieto, S. & Garmendia, J. (1998). Spatial and temporal patterns of annual precipitation variability over the Iberian Peninsula. *International Journal of Climatology*, 18, 299-316.

Rogers, J. (1984). The association between the North Atlantic oscillation and the southern oscillation in the Northern Hemisphere. *Monthly Weather Review*, 112, 1999-2015.

Serrano, L., Antunes, S. & Oliveira Pires, H. (2003). Temperature and precipitation spatial variability in Portugal. *Proceedings of 3rd Symposium on Meteorology and Geophysics*, Associação Portuguesa de Meteorologia e Geofísica, ISBN 972-99276-0-X, Aveiro, Portugal (in Portuguese).

Stephenson, D., Pavan, V. & Bojariu, R. (2000). Is the North Atlantic oscillation a random walk? *International Journal of Climatology*, 20, 1–18.

Trigo, R., Pozo-Vázquez, D., Osborn, T., Castro-Díez, Y., Gámiz-Fortis, S. & Estaban-Parra, M. (2004). North Atlantic Oscillation influence on precipitation, river flow and water resources in the Iberian Peninsula. *International Journal of Climatology*, 24, 925-944.

Ulbrich, U., Cristoph, M., Pinto, J. & Corte-Real, J. (1999). Dependence of winter precipitation over Portugal on NAO and baroclinic wave activity. *International Journal of Climatology*, 19, 379-390.

Vautard, R. (1995). *Analysis of climate variability – applications of statistical techniques*. von Storch H., Navarra A. (Eds.). Springer.

Vautard, R., Yiou, P. & Ghil, M. (1992). Singular spectrum analysis: A toolkit for short, noisy chaotic signals. *Physica D*, 58, 95-126.

von Storch, H., Zorita, E., Jones, J., Dimitriev, Y., González-Rouco, F. & Tett, S. (2004). Reconstructing past climate from noisy proxy data. *Science*, 306, 679–682.

von Storch, H. & Zwiers, F. (1999). *Statistical Analysis in Climate Research*. Cambridge University Press, ISBN 0 521 45071 3, Cambridge, United Kingdom.

Wanner, H., Beer, J., Bütikofer, J., Crowley, T., Cubasch, U., Flückiger, J., Goosse, H., Grosjean, M., Joos, F., Kaplan, J., Küttel, M., Müller, S., Pentice, C., Solomina, O., Stocker, T., Tarasov, P., Wagner, M. & Widmann, M. (2008). Mid to late Holocene climate change – an overview. *Quaternary Science Reviews*, 27, 1791–1828.

Zorita, E., Kharin, V. & von Storch, H. (1992). The atmospheric circulation and sea surface temperature in the North Atlantic area in winter: their interaction and relevance for Iberian precipitation. *Journal of Climate*, 5, 1097-1108.

10

Application of Principal Components Regression for Analysis of X-Ray Diffraction Images of Wood

Joshua C. Bowden[1] and Robert Evans[2]
[1]*CSIRO Information Management and Technology,*
[2]*CSIRO Materials Science and Engineering,*
Australia

1. Introduction

We report on the use of Principal Component Regression (PCR), a least squares calibration method employing Principal Component Analysis (PCA) for the analysis of large sets of images obtained by wide angle X-ray diffractometry (WAXD) of wood. PCR multivariate models were created for prediction of microfibril angle (MFA), an important wood structural parameter. Multivariate regression techniques have previously been applied to X-ray diffraction data for a variety of analytical problems. These include analysis of complex mineral mixtures (Karstang and Eastgate 1987) and for automated high-throughput analysis of powder diffraction profiles obtained by azimuthal averaging of 2D diffraction patterns (Gilmore, Barr et al. 2004). Multivariate classification from X-ray diffraction data has also been used for discrimination of insulin microcrystals (Norrman, Stahl et al. 2006) and brucite particle morphology (Matos, Xavier et al. 2007). The models created in this study used pixels from across the full area of the diffraction image as predictor variables, a technique which has not been found applied to diffraction data in the literature. A related technique, multivariate analysis of congruent images (MACI), was introduced by Eriksson, Wold et al. (2005) for discrimination and quantification of events within video image sequences. MACI uses wavelet analysis to decompose images and obtain predictive information which is correlated with frame-by-frame differences and that can describe properties such as the movement of an object or change in composition of a feature.

This chapter describes four predictive PCR models, two that are species-specific, and two that incorporate multiple species. Differences between these models, and their relative accuracy, will be discussed. SilviScan will be described and the issues relating to management of the large amounts of data collected by the system will be examined.

2. Multivariate calibration

Principal component regression is a widely used multivariate calibration technique for creating predictive models by relating multivariate sets of measurements ($\mathbf{X}$) to sets of dependent variables ($\mathbf{y}$) (Hoskuldsson 1995). PCR uses the properties of principal component analysis to reduce the dimensionality of the original data to form linearly independent factors (T) required for least squares regression analysis. In the case of X-ray diffraction analysis, modelling of a dependent variable using PCR relies on the the proportionality that exists between diffracted X-ray energy and the quantity of a compound or structural feature present within a sample. If there is a linear relationship with the predictand, then PCR will be a good choice for modelling the variable using X-ray diffraction images. A linear relationship between a dependent variable and diffraction data is valid providing the diffraction experimental parameters are within prescribed limits of a materials density (these limits are dependent on X-ray wavelength) and providing all other experimental conditions are kept constant (Tripp and Conrad 1972; Cave and Robinson 1998).

PCR involves three main steps. Step 1 is data reduction using PCA to obtain scores (T) and normalised eigenvectors (P) (Equation 1 and 2). Step 2 uses T as the independent variables matrix in the inverse least squares (ILS) calibration (Kramer 1998) (Equation 3 and 4), and step 3 produces the regression coefficients (in terms of the original X variables) required for predicting the properties (ŷ) of unknown samples (Equation 5). PCA is represented by Equation 1 (Brereton 2003). The eigenvector and score pairs are termed principal components (PCs) and are numbered from PC1 to PCn, in decreasing order of the variance associated with the vectors in the original data matrix. The matrix R is the residual data matrix which is the variance not associated with the principal components chosen, and which would generally describe noise. Equation 2 represents the projection of the eigenvectors onto the original data to obtain the reduced data (scores) T matrix:

$$\mathbf{X}^{cal}_{m,n} = \mathbf{T}_{m,c}\mathbf{P}_{c,n} + \mathbf{R}_{m,n} \tag{1}$$

$$\mathbf{T}_{m,c} = \mathbf{X}^{cal}_{m,n}\mathbf{P}'_{c,n} \tag{2}$$

where m is the number of samples, n is the number of variables - in our case the total number of pixels in the image, c is the number of components and ' indicates the transpose. For analysis involving images, the term 'eigenimage' is used to denote the 2D version of the eigenvector that is obtained from PCA.

The equations required for the inverse least squares regression calibration step are:

$$\mathbf{y}'_{m,1} = \mathbf{b}_{1,c}\mathbf{T}'_{m,c} \tag{3}$$

$$\mathbf{y}'_{m,1}\cdot\mathbf{T}\left[\mathbf{T}'\mathbf{T}\right]^{-1} = \hat{\mathbf{b}}_{1,c} \tag{4}$$

$$\boldsymbol{\beta}_{1,n} = \hat{\mathbf{b}}_{1,c}\mathbf{P}_{c,n} \tag{5}$$

$$\hat{\mathbf{y}}_{x,1} = \mathbf{X}_{x,n}\boldsymbol{\beta}'_{1,n} + \bar{y}_{est} \tag{6}$$

where $y_{m,1}$ is the vector of dependent variables, b and $\hat{b}_{1,c}$ are the true and best estimates for the least squares coefficients, $\beta_{1,n}$ are the regression coefficients in terms of the original X variables, $\hat{\mathbf{y}}$ are the predicted (best estimates of) $\mathbf{y}$ values from any other set of data and $\bar{y}_{est}$ is the best estimation of the average $\mathbf{y}$ data.

The ILS calibration requires that the number of factors used (c) is less than or equal to the number of samples (m). Once the calibration is performed, $\mathbf{y}$ data may be predicted from unknown samples ($\mathbf{X}$) using the regression coefficients β of the original $\mathbf{X}$ variables and the best estimation ($\bar{y}_{est}$) of the average $\mathbf{y}$ data. The $\bar{y}_{est}$ value can be obtained using an added column vector of 1's in the $\mathbf{T}$ matrix and will equal the average of the $\mathbf{y}$ dataif the $\mathbf{X}$ matrix was mean centred prior to the PCA step.

The scores obtained from PCA are orthogonal and rule out modelling techniques that use cross-terms ($\mathbf{T}_{m,c1} \times \mathbf{T}_{m,c2}$). However, both cross and quadratic ($\mathbf{X}_{m,n} \times \mathbf{X}_{m,n}$) terms may be introduced into the initial $\mathbf{X}$ data, which in the case of large data sets comes at the risk of excessive memory use. Variables can be scaled prior to the PCA step by one of a number of techniques such as mean centering (subtracting the mean of a variable), variance scaling (equalising variance) and range scaling (normalising data to lie within a nominated range). Scaling techniques are used to improve the predictive capability of the data (Berg, 2006) and can highlight low variance variables that may otherwise be left in the $\mathbf{R}$ matrix during the PCA step. Vector normalisation (VN) is another scaling technique that normalises between samples (rather than variables) and involves the following transformation:

$$\mathbf{x}_{vn} = \frac{\mathbf{x}}{\sqrt{\sum_{i=1}^{n} x_i^2}} \qquad (7)$$

where $\mathbf{x}_{vn}$ is the normalised vector, $\mathbf{x}$ the original vector and x_i the ith variable of the sample.

$$\mathbf{H} = \mathbf{X}\left[\mathbf{X}'\mathbf{X}\right]^{-1}\mathbf{X}' \qquad (8)$$

Removal of high leverage samples from the model data set before inclusion in the model training data is good practice as they influence the model in a disproportionate fashion. The leverages for each sample are the values of the diagonal of the 'hat' matrix ($\mathbf{H}$) as seen in equation 8 (Dobson, 2002).

A useful measure of the accuracy of a multivariate regression model is the root mean square standard error of prediction (RMSEP) (Faber, Song et al. 2003):

$$\text{RMSEP} = \sqrt{\frac{\sum_{i=1}^{m}\left(y_i - \hat{y}_i\right)^2}{m}} \qquad (9)$$

and is a measure of the average error between a test set of reference samples and a models predicted values ($\hat{y}$). This measure is biased high when using imprecise reference values (y) and can be corrected if the standard deviation of the reference data is known (Bro, Rinnan and Faber, 2005). Tests are available for determination of the per-sample prediction error, based on the leverage of the sample (see Bro, Rinnan and Faber, 2005).

The optimum number of components to keep in a model can be determined from the minimum in RMSEP versus the number of components (Martens and Naes 1989). Another common measure of prediction error is RMSEP from cross calibration. Cross calibration requires generation of multiple models with exclusion of a small number of samples (usually one) from each model and calculating the error as the average of the RMSEP of the excluded samples for all models. This technique is not practical or necessary when data sets are large and in our case where there would be a large number of models to be made (each potentially requiring PCA on the subset of data), the computational requirements would become excessive.

3. X-ray diffraction of wood and the SilviScan system

Important factors affecting the diffraction patterns of wood samples include cell shape and microfibril angle (MFA) (Evans 1999; Lichtenegger, Reiterer et al. 2001), cellulose crystallite dimensions and cellulose crystallinity (Wada, Okano et al. 1997; Washusen and Evans 2001; Andersson, Serimaa et al. 2003; Garvey, Parker et al. 2005). These properties have been shown to influence many of the bulk properties of wood including modulus of elasticity (MoE) for structural timber (Evans 2006) and important properties such as pulp yield, a useful metric for production capacity in the pulping industry (Downes, Evans et al. 2003). The fast and accurate estimation of these attributes is useful in large scale surveys of wood properties for genetic selection trials and forestry management (Lindstrom, Harris et al. 2004; Thumma, Nolan et al. 2005).

Wide angle x-ray diffractometry has been developed as an integral part of the high-throughput instrumentation known as SilviScan, now in its 3rd generation. SilviScan consists of three complimentary techniques, transverse (radial-tangential) surface microscopy and image analysis, X-ray densitometry and WAXD. These three techniques are used to analyse 2mm by 7mm radial sections, typically cut from 12mm diameter increment cores taken from standing trees (Fig. 1).

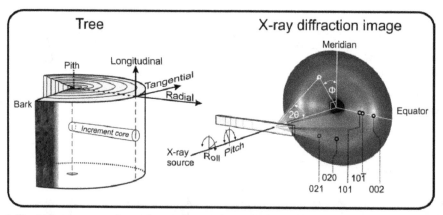

Fig. 1. Typical geometry of wood samples used in relation to the tree and the diffraction pattern produced. Miller indices associated with the observed reflections are indicated. The commonly used 040 meridional reflection is outside the range of the camera and is unreliable when fibre orientation varies as it does in core samples.

The interaction of an X-ray beam with the wood fibre structure gives diffraction images that contain information about the helical orientation of the fibres (MFA). A geometrical relationship between MFA and the variance of the observed 002 azimuthal peaks (Φ direction, Fig.1) is currently used by SilviScan software to determine MFA (Evans 1999). In addition, 3D cellular orientation is estimated (roll and pitch relative to the x-ray beam). Roll is calculated from the rotation of the diffraction pattern around the beam direction and pitch is calculated from the distortion of equatorial reflections towards the meridional axis. Pitch (fibre tilt in the direction of the X-ray beam) is important as it causes asymmetric peak broadening and an overestimation of MFA (Stuart and Evans 1995; Evans, Stringer et al. 2000), which has to be corrected for during SilviScan analysis.

This work demonstrates the applicability of multivariate statistical calibration methods for determination of MFA from WAXD images obtained on SilviScan. The same method can be used to extract information on fibre orientation, crystallite size and any other property that affects the images. Occasionally, when MFA is very high, the diffraction patterns have almost no azimuthal intensity variation, making them very difficult to analyse by any method that relies on the determination of azimuthal peak widths. The consequences of this problem are data 'dropouts' and an example of them are seen later in the results section. Thus overcoming this problem through calibrated multivariate techniques for estimation of MFA is of interest.

4. Material and methods

4.1 SilviScan

The SilviScan 3 instrument scans samples in three stages. First, automated microscopy and image analysis of the polished top transverse surface (radial-tangential plane) gives growth-ring boundary orientation, ray cell orientation, and average radial and tangential fibre diameters binned in 25 micron intervals. Second, automated X-ray scanning densitometry (customised 1392×1040 pixel camera, Photonic Science, UK) gives density variation binned in 25 micron intervals and normalised to the actual average sample density obtained by independent gravimetric analysis. Third, automated scanning X-ray diffractometry (customised 1392×1040 pixel intensified camera, Photonic Science, UK) using a nickel-filtered Cu source and 200 micron diameter capillary-focussed beam, gives microfibril angle (MFA) and fibre axis orientation in 100 micron steps. MFA was corrected for fibre orientation in the beam direction.

4.2 Wood samples

All wood samples (two softwoods and one hardwood) were taken from previous projects for which the SilviScan data was available. The samples were in the form of 2mm wide (tangential dimension) and 7mm high (longitudinal dimension) pith-to-bark (radial) strips. Samples varied in length from 20 mm – 200 mm.

30-year-old *Pinus caribaea* var. *hondurensis* (*P. caribaea*) from Beerburrum, in south-east Queensland, Australia. Twelve radial samples from an original 520 (65 trees, 8 heights per tree) were selected for high resolution analysis on SilviScan-3 at the 4m sampling height. 18486 diffraction images were taken of these samples, 10000 used for creating the models and 6925 images were used for model verification.

23-year-old *Pinus radiata* (*P .radiata*) from Tallaganda, Canberra, Australia were taken from a set of previously well characterised samples as 2mm tangential, 7 mm longitudinal and 20 mm radial strips (Schimleck 2002). The resulting diffraction dataset consisted of 11822 images from 116 paired ends of the same sample.

Eucalyptus delegatensis (*E. delegatensis*) from East Gippsland, Victoria, Australia were taken as an example of a hardwood (Evans and Ilic 2001). The resulting diffraction image dataset consisted of 10638 images from 102 paired ends of the same sample.

4.3 Data analysis

Principal component analysis was undertaken using in-house software developed using C++ and OpenCL and making use of high end graphics processing units (Bowden 2010). Images were reduced to 128 by 128 pixels by selecting 1 out of 8 pixels in each dimension. This resulted in a 16384 element vector for each image in the data sets. Inverse least squares software was written in-house and the combined system used hierarchical data format (HDF) files for data storage and management. HDF Viewer version 2.7 and Microsoft Excel 2003 was used for visualisation and data filtering (HDF 2000).

The modelling procedures involved the following: Two datasets were used to create the models, one from the *P. caribaea* data set (10000 images, ~7 wood samples) and another from a combination of *P. radiata* (5000 images), *E. delegatensis* (5000 images) and *P. caribaea* (10000 images), making a 20000 image data set. These are referred to as the 'single-species' and 'multi-species' data sets. From each data set, models were made using mean centred image data of both vector normalised (referred to as 'VN' models) and unmodified data. Models were then tested with 6925 *P. caribaea* and 5000 samples of either *P. radiata* or *E. delegatensis*. RMSEP model verification was carried out on a per species basis with models from both data sets.

5. Results

Fig. 2 shows two diffraction images at opposite ends of the range of MFA values seen in *P. caribaea* samples. Fig. 2a is a low MFA sample (9°), with very narrow azimuthal peaks (101, 10$\bar{1}$ and 002 in order from the centre) close to the equator. Fig. 2b is an example of a high MFA sample (30°) and the azimuthal peaks are much broader. The arrows (Fig. 2a) on the 002 reflections indicate the direction of azimuthal broadening in high MFA samples.

Fig. 3 shows the first 18 eigenvectors/eigenimages from PCA of the single-species dataset. The eigenimages show various symmetry features involving rotation, reflection and inversion. These symmetry features are characteristics of the diffraction pattern derived from wood samples and result from the major sources of variance in the diffraction images. For example, the first eigenimage can be interpreted as the primary contribution to variation in equatorial intensity with complementary variation in MFA (azimuthal peak width).

The eigenimage PC2 arises from differential absorption of equatorially scattered X-rays. This differential is caused mainly by rotation of the samples in the horizontal plane and strong horizontal variations in wood density (annual rings) in the radial direction. PC3 is the result of changes due to tilting (roll) of the fibre around the X-ray beam, causing the main diffraction peaks to be rotated away from the equator and described by rotational symmetry combined with a lack of reflection symmetry. Some higher order PCs such as PC7 do not show rotational symmetry and probably arise from fibre tilt in the direction of the X-ray beam (pitch) and rotation around the fibre axis.

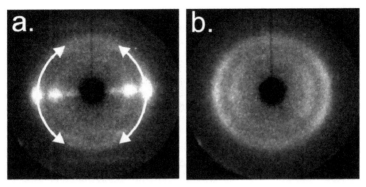

Fig. 2. Example of a. low (9°) and b. high (30°) MFA diffraction images from the *P. caribaea* samples.

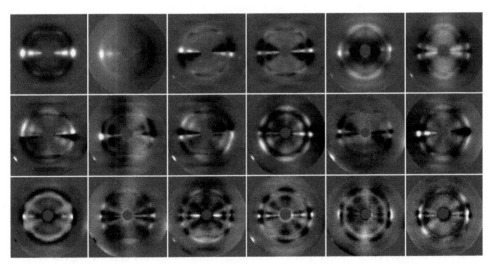

Fig. 3. The first 18 eigenimages of 10000 vector normalised diffraction images obtained from the single-species dataset. PCs are in numeric order going left to right, top to bottom. Bright areas indicate positive values, dark areas negative and mid-grey zero (The absolute sign of the pixels is irrelevant). These images are the 'unit', unweighted components of the modelled image.

Fig. 4 shows a logarithmic plot of eigenvalue (the dot product of each scores vector with itself) for successive PCs from the mean-centred, vector-normalised single-species data. The eigenvalues rapidly decrease by over 3 orders of magnitude.

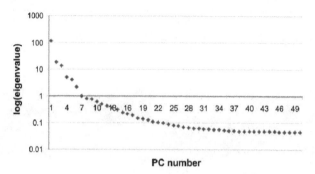

Fig. 4. Eigenvalue for each PC for the mean-centred, vector-normalised single-species PCA model.

As discussed in section 2 above, the ILS modelling procedure incorporates the eigenimage vectors into a predictive model of the dependent variable using the proportionality found within the **T** matrix (Equation 4). The regression coefficients (β) such as those seen in Fig. 5 are also generated. These coefficients show a distinct pattern of strongly negative values at the equatorial positions and strongly positive values moving into the equatorial-meridian arc.

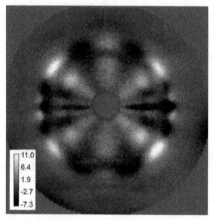

Fig. 5. Regression coefficients obtained from the single-species model using 17 PCs. The image shows distinct positive bands placed diagonally across the image and extending to the limit of the 002 peak 2θ value. For these images both the intensity and the sign of the values is important.

This pattern follows the behaviour of the original data (Fig. 2), where high MFA samples exhibit broad, low intensity equatorial peaks and low MFA samples show more intense, narrow peaks. Thus the model can predict the MFA of unknown samples by summing the product of each pixel value in the unknown image and the regression coefficient at that pixel position, and adding the best estimate for the mean MFA of the calibration set. It is understood that the unknown image is collected on a similar X-ray camera, with matching effective experimental geometry and pre-processed in the same way as the calibration images.

Fig. 6 shows that the RMSEP of the *P. caribaea* test data is reduced only when specific eigenimages are added to the model. The contributing PCs are seen to be PC1, 4, 5, 6, 10, 14, 16, and 17. When referenced against the eigenimages in Fig. 3, the influential PCs are seen to have a dominant characteristic of two orthogonal mirror planes, one through the meridian and another at the equator. This symmetry arises from the vertically aligned wood fibres and the horizontal scattering of x-rays from the crystalline cellulose c-axis planes which, on average are aligned with the fibre axis.

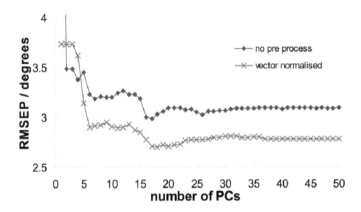

Fig. 6. RMSEP derived from the 10000 image single-species model using 6925 *P. caribaea* verification samples shows that vector normalisation gives consistently better prediction than non-vector normalised (mean centred only) data. The vector normalised samples reach a minimum on the addition of PC17 to the model.

Fig. 7 shows the scatter plots of predicted MFA versus the reference data MFA for the three species as obtained using the single-species data set. It is seen that the single species model predicts *P. caribaea* reference data well, with the predicted data following the line of perfect correlation adequately (RMSEP = 2.7° at 17 PCs for VN samples). The prediction of *P. radiata* samples is also found to be accurate (RMSEP = 1.7° using 19 PCs for the VN samples), with a deviation in the slope producing an overestimation of high MFA samples.

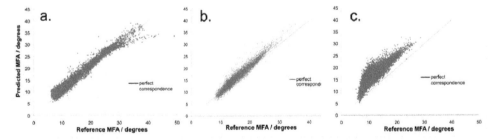

Fig. 7. Reference versus predicted scatter plots of MFA from the single-species model using 17 PCs and VN samples for the predictive model. a. *P. caribaea*, b. *P. radiata* and c. *E. delegatensis* reference samples.

The *E. delegatensis* samples do not fair so well, with a consistent shift in the predicted MFA values over the reference values (RMSEP = 5.4° using 37 PCs in the VN samples). Analysis shows that much of the error in the *E. delegatensis* samples is due to the $\bar{y}$ value obtained from the *P. caribaea* (17.1°) not being a good estimate for the *E. delegatensis* $\bar{y}$ value (11.9°). Correcting for this difference and using it in the model (i.e. offsetting the *E. delegatensis* predicted data by 17.1-11.9) results in a reasonable RMSEP of 2.1° using 18 PCs.

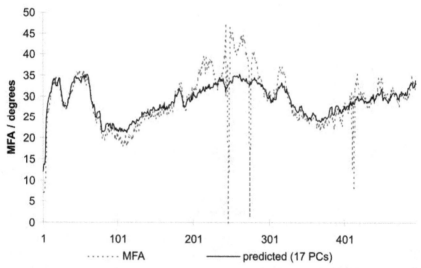

Fig. 8. Plot of predicted MFA versus the reference MFA data for a selected region of high MFA within the early growth wood of a *P. caribaea* sample. Model is from 17 PCs using the VN single-species data only. X axis represents sample number moving along a wood sample.

The data in Fig. 8 shows the advantage of using models such as these as a verification tool for data measured by other procedures. Three data 'dropouts' can be observed in a region of high MFA in the reference MFA data in Fig. 8. These problems are possibly due to high fibre pitch, which distorts the diffraction patterns and thus cause the direct determination of MFA to fail occasionally. It is seen however, that the multivariate model has predicted MFA to follow a profile under the reference data and does not have the problem of dropouts.

Fig. 9 shows that improvements can be made to the predictive potential of models by using multiple species in the calibration data. Fig. 9a shows that the multi-species model marginally increases the accuracy of prediction of *P. radiata* samples, notably drawing the high MFA samples onto the line of perfect correspondence (RMSEP = 1.2° using 37 PCs, VN samples). Fig. 9b shows that the *E. delegatensis* samples are still not well modelled (RMSEP = 4.2° using 18 PCs). The offset is predominately due to the difference in the average MFA between the multi-species dataset (15.2°) and the *E. delegatensis* reference samples (12.1°). This lack of fit indicates that the average values used in offsets of the models must be carefully monitored and that it may be necessary to generate species specific models.

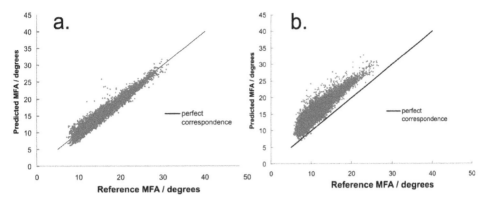

Fig. 9. Reference versus predicted scatter plots of MFA being predicted from the 200000 image multi-species model a. Prediction of *P. radiata* samples MFA from model using 37 PCs and b. Prediction of *E. delegatensis* samples using 17 PCs in the model.

6. Discussion

The PCA modelling procedure reduced the dimensionality of the image dataset from the 16384 pixels present in the original images to a sum of contributions of a smaller number of principal components. Each component found had a characteristic (and mutually orthogonal) distribution of intensities over the 2D image plane. The PCA technique selects the features of these components in order of contributing variance to the (pre-processed) data set. Due to the broad nature of the diffraction features in wood samples, PCA was able to obtain PCs that describe most of the variance in the images. The inverse least squares procedure then used the newly formed variables as predictors to produce a least squares model which can accurately predict MFA. Vector normalised image data produced more accurate models in all experiments than unprocessed, mean centred only data. An R^2 of 0.96 was obtained for *P. caribaea* samples in both single-species and mixed-species models. *P. radiata* R^2 reached a maximum around 0.93 for both models and *E. delegatensis* R^2 increased from 0.84 to 0.91 when the model was produced from the multi-species data set.

Although normalisation procedures are effective, use of regression coefficients with data from other X-ray systems will suffer from errors due to between system differences. These problems are typical of PCR modelling techniques that are sensitive to the precise instrumental system being calibrated.

A disadvantage of this analysis method for large image datasets is the computer memory requirements needed in creation of the models. A single wood sample could generate up to 3000 1.4 Mpixel images, and there may be hundreds of such samples in a project. These difficulties can be overcome through use of high performance cluster-based computing systems and the use of graphics processing units (GPUs) to speed up the PCA data reduction procedure. In the analysis above, images were down-sized to 128 by 128 pixels by discarding pixels. Although not shown, the use of more pixels in the analysis or use of averaging of the pixels should improve the resulting accuracy of models produced. The data analysis system found benefit from use of the HDF data format as it allowed both data and results to be grouped, with each experiment recorded and similar experiments able to be filed together. The HDF format allows data importation into multiple other popular data analysis programs and thus aides data analysis by researchers with differing software access.

The computational burden of the final prediction part of the procedure is relatively low compared with mapping Φ-θ images into orthogonal axes and segmentation of the image as is presently used by the geometric method of MFA prediction within the SilviScan system. Thus as a technique to increase the robustness of a high throughput system, whole image multivariate property prediction has potential.

A further advantage of multivariate techniques is that standard methods for outlier detection can be used and thus samples that lie outside the calibration can be highlighted and scrutinized by an operator. The ability to emphasize unusual samples is of interest in a high throughput system, where the amount of data collected often precludes a thorough assessment of all results obtained. The technique should tolerate noisy images and allow faster sample throughput.

7. Conclusions

The calibration techniques used here open the way for rapid determination of difficult to obtain wood properties. Measurements from instruments that can accurately measure properties of interest, albeit more slowly can be used for calibration and subsequently used in the high throughput instrument. Models will have to be carefully designed and tested such that they will be valid for a species of interest. These models can be incorporated into the instrument software for checking data integrity. In addition, the signal-to-noise advantage inherent in employing the whole image as a predictor could be used to obtain accurate values at a faster rate.

8. Acknowledgment

Forestry Plantations Queensland are gratefully acknowledged for use of *P. caribaea* samples. Dr F. Chen and Dr M. Hilder for reading of manuscript in early stages, Ms S. Harper for practical help on the SilviScan system and Mr N. Ebdon for running samples.

9. References

Andersson, S., Serimaa R., Paakkari, T., Saranpaa, P., Pesonen, E. (2003). Crystallinity of wood and the size of cellulose crystallites in Norway spruce (*Picea abies*). *Journal of Wood Science*, Vol. 49, No. 6, pp.531-537.

Berg, R. A. V. D., Hoefsloot, H. C., Westerhuis, J. A., Smilde, A. K. and Werf, M. J. V. D. (2006). Centering, scaling, and transformations: improving the biological information content of metabolomics data. *BMC Genomics*, Vol. 7, No. 142.

Bowden, J. C. (2010). Application of the OpenCL API for Implementation of the NIPALS Algorithm for Principal Component Analysis of Large Data Sets. e-Science Workshops, 2010 Sixth IEEE International Conference on e-Science.

Brereton, R. G. (2003). *Chemometrics: data analysis for the laboratory and chemical plant.* Wiley, 9780471489788, Chichester, West Sussex, England; Hoboken, NJ.

Bro, R., Rinnan, A., Faber, N.M. (2005). Standard error of prediction for multilinear PLS2. Practical implementation in fluorescence spectroscopy. *Chemometrics and Intelligent Laboratory Systems* Vol. 75, pp. 69 – 76

Cave, I. D. and Robinson, W. (1998). Measuring microfibril angle distribution in the cell wall by means of x-ray diffraction. *Microfibril angle in wood*. B. G. Butterfield. pp. 94-107, Christchurch, University of Canterbury.

Downes, G., Evans, R., Wimmer, R., French, J., Farrington, A. and Lock, P. (2003). Wood, pulp and handsheet relationships in plantation grown *Eucalyptus globulus*. *Appita Journal*, Vol. 56, No. 3, pp. 221-228.

Dobson, A.J. (2002) An introduction to generalized linear models 2nd ed. Chapman and Hall/CRC, 1584889500 Boca Raton, Florida.

Eriksson, L., Wold, S. and Trygg, J. (2005). Multivariate analysis of congruent images (MACI). *Journal of Chemometrics*, Vol. 19, No. 5-7, pp. 393-403.

Evans, R. (1999). A variance approach to the X-ray diffractometric estimation of microfibril angle in wood. *Appita Journal*, Vol. 52, No. 4, pp. 283.

Evans, R., Stringer, S. and Kibblewhite, R. P. (2000). Variation of microfibril angle, density and fibre orientation in twenty-nine *Eucalyptus nitens* trees. *Appita Journal*, Vol. 53, No. 6, pp. 450-457.

Evans, R. and Ilic, J. (2001). Rapid prediction of wood stiffness from microfibril, angle and density. *Forest Products Journal*, Vol. 51, No. 3, pp. 53-57.

Evans, R. (2006). Wood stiffness by X-ray diffractometry. *Charcterization of the cellulosic cell wall*. D. D. Stokke and L. H. Groom. Iowa, Blackwell Publishing.

Faber, N. M., Song, X. H. and Hopke, P. K. (2003). Sample-specific standard error of prediction for partial least squares regression. *TrAC Trends in Analytical Chemistry*, Vol. 22, No. 5, pp. 330-334.

Garvey, C. J., Parker, I. H. and Simon, G. P. (2005). On the interpretation of X-ray diffraction powder patterns in terms of the nanostructure of cellulose I fibres. *Macromolecular Chemistry and Physics*, Vol. 206, No. 15, pp. 1568-1575.

Geladi, P. and Kowalski, B. R. (1986). Partial least squares regression – a tutorial. *Analytica Chimica Acta*, Vol. 185, pp. 1-17.

Gilmore, C. J., Barr, G. and Paisley, J. (2004). High-throughput powder diffraction. I. A new approach to qualitative and quantitative powder diffraction pattern analysis using full pattern profiles. *Journal of Applied Crystallography*, Vol. 37, No. 2, pp. 231-242.

The HDF Group. Hierarchical data format version 5, 2000-2010. http://www.hdfgroup.org/HDF5

Hoskuldsson, A. (1995). A combined theory for PCA and PLS. *Journal of Chemometrics*, Vol. 9, No. 2, pp.91-123.

Karstang, T. V. and Eastgate, R. J. (1987). Multivariate calibration of an x-ray diffractometer by partial least squares regression. *Chemometrics and Intelligent Laboratory Systems*, Vol. 2, No. 1-3, pp.209-219.

Kramer, R. (1998). *Chemometric Techniques for Quantitative Analysis*. Marcel Dekker, Inc., 9780824701987, New York.

Lichtenegger, H. C., Reiterer, A., Stanzl-Tschegg, S. E. and Fratzl, P. (2001). Comment about The measurement of the micro-fibril angle in soft-wood by K. M. Entwistle and N. J. Terrill. *Journal of Materials Science Letters*, Vol. 20, No. 24, pp.2245-2247.

Lindstrom, H., Harris, P., Sorensson, C. T. and Evans, R. (2004). Stiffness and wood variation of 3-year old *Pinus radiata* clones. *Wood Science and Technology*, Vol. 38, No. 8, pp.579-597.

Matos, C. R. S., Xavier, M. J., Barreto, L. S., Costa, N. B. and Gimenez, I. F. (2007). Principal component analysis of X-ray diffraction patterns to yield morphological classification of brucite particles. *Analytical Chemistry*, Vol. 79, No. 5, pp.2091-2095.

Martens, H. and Naes, T. (1989) *Multivariate calibration*. Wiley, 0471909793 Chichester [England]; New York.

Norrman, M., Stahl, K., Schluckebier, G. and Al-Karadaghi, S. (2006). Characterization of insulin microcrystals using powder diffraction and multivariate data analysis. *Journal of Applied Crystallography*, Vol. 39, No. 3, pp.391-400.

Schimleck, L. R., Evans, R. and Matheson, A. C. (2002). Estimation of Pinus radiata D. Don clear wood properties by near-infrared spectroscopy. *Journal of Wood Science* Vol. 48, No. 2, pp.132-137.

Stuart, S-A and Evans, R. (1995) Effect of fibre tilt on the estimation of the microfibril angle by x-ray diffraction. 49th Appita Annual General Conference, April 3-7, Hobart, Tasmania, Australia.

Thumma, B. R., Nolan, M. R., Evans, R. and Moran, G. F. (2005). Polymorphisms in cinnamoyl CoA reductase (CCR) are associated with variation in microfibril angle in Eucalyptus spp. *Genetics*, Vol. 171, No. 3, pp.1257-1265.

Tripp, V. W. and Conrad, C. M. (1972). X-Ray Diffraction. *Instrumental analysis of cotton cellulose and modified cotton cellulose*. R. T. O'Connor. Marcel Dekker, 9780824715007, New York.

Wada, M., Okano, T. and Sugiyama, J. (1997). Synchrotron-radiated X-ray and neutron diffraction study of native cellulose. *Cellulose*, Vol. 4, No. 3, pp.221-232.

Washusen, R. and Evans, R. (2001). Prediction of wood tangential shrinkage from cellulose crystallite width and density in one 11-year-old tree of *Eucalyptus globulus* Labill. *Australian Forestry*, Vol. 64, No. 2, pp.123-126.

11

Automatic Target Recognition Based on SAR Images and Two-Stage 2DPCA Features

Liping Hu[1], Hongwei Liu[2] and Hongcheng Yin[1]

[1]National Key Laboratory of Target and Environment Electromagnetic Scattering and Radiation, Beijing Institute of Environmental Characteristics, Beijing,
[2]National Key Laboratory of Radar Signal Processing, Xidian University, Xi'an,
China

1. Introduction

In recent years, radar Automatic Target Recognition (ATR) based on target synthetic aperture radar (SAR) images has received more and more attentions. So far, many literatures based on MSTAR public dataset are released, which focus on the SAR target recognition related techniques including target segmentation, feature extraction, classifier design, and so on. A template matching was proposed (Ross et al., 1998). Support Vector Machine (SVM) has been applied to SAR ATR (Zhao & Principe, 2001; Bryant & Garber, 1999). The drawbacks of them are that none of them have any pre-processing and feature extraction. However, efficient pre-processing and feature extraction may help to improve recognition performance.

Principal Component Analysis (PCA) is a classical feature extraction technique. But when PCA is used for images feature extraction, 2D image matrices must be previously transformed into 1D image vectors. This usually leads to a high dimensional vector space, where it is difficult to evaluate the covariance matrix accurately. To solve this problem, 2-dimensional PCA (2DPCA) for image feature extraction is proposed (Yang et al., 2004). As opposed to PCA, 2DPCA constructs the covariance matrix directly using 2D image matrices rather than 1D vectors, and evaluates the covariance matrix more accurately. Moreover, the size of the covariance matrix is much smaller. A drawback of 2DPCA is that it only eliminates the correlations between rows. So it needs more features, and this will lead to large storage requirements and cost more time in classification phase. To further compress dimension of features, two-stage 2DPCA is applied in this chapter.

The remainder of this chapter is organized as follows: in Section 2, the SAR images pre-processing method is described. 2DPCA is first reviewed, and two-stage 2DPCA is described in Section 3. In Section 4, classifiers are described. In Section 5 and 6, experimental results based on Moving and Stationary Target Acquisition and Recognition (MSTAR) data and conclusions are presented.

2. SAR image pre-processing

The original SAR images provided by MSTAR contain not only targets of our interest, but also background clutters, as shown in Fig. 1 (a). If targets are recognized based the original images, clutters would depress the system performance. Thus, it is necessary to pre-process the original images to segment targets from background clutters.

2.1 Logarithmic transformation

We transform the original images using logarithm conversion, which can convert speckles from multiple model to additional model and make the image histogram more suitable be approximated with a Gaussian distribution. The logarithmic transformation is given by

$$G(x,y) = 10 \lg \left[F(x,y) + 0.001 \right] + 30 \tag{1}$$

where F denotes the magnitude matrix of the original SAR image. Since the logarithm is not defined at 0, we add an arbitrary constant (for example 0.001) to the original image before the logarithm. To ensure the pixel values to be nonnegative, we add a corresponding constant (30).

2.2 Adaptive threshold segmentation

In order to obtain the target image, the adaptive threshold segmentation method is adopted. First of all, estimating the mean μ and the variance σ of the current image G, for each pixel (x,y) of G

$$\begin{cases} (x,y) \in T_{ar}, T_{ar}(x,y) = 1, & if \;\; G(x,y) > \mu + c\sigma \\ (x,y) \in B_{ac}, T_{ar}(x,y) = 0, & else \end{cases} \tag{2}$$

Where T_{ar}, B_{ac} denote the target and the background respectively, c can be obtained statistically from training samples.

2.3 Morphological filter and geometric clustering operation

Due to the presence of speckles, the result of threshold segmentation contains not only target, but smaller objects inevitably, as shown in Fig. 1 (b). To remove these small objects and obtain smoothing the target image, morphological filter (Gonzalez &Woods, 2002) and geometric clustering operation (Musman & Kerr, 1996) are adopted to T_{ar}.

Morphological filter aims to smooth boundary, remove sharp protrusions, fill small concaves, remove small holes, joint gaps, and so on.

In general, filtered image T_{ar} may also contain some non-target regions, which are much smaller than target itself, as shown in Fig. 1 (c). To remove small regions, we apply geometric clustering operation: firstly, detect and label all the independent connected regions in T_{ar}. Then, compute areas for each region. The largest region is of our interest. In this way, we obtain the resulting T_{ar}, as shown in Fig. 1 (d). Overlaying the resulting T_{ar} on the logarithmic image G obtains target intensity.

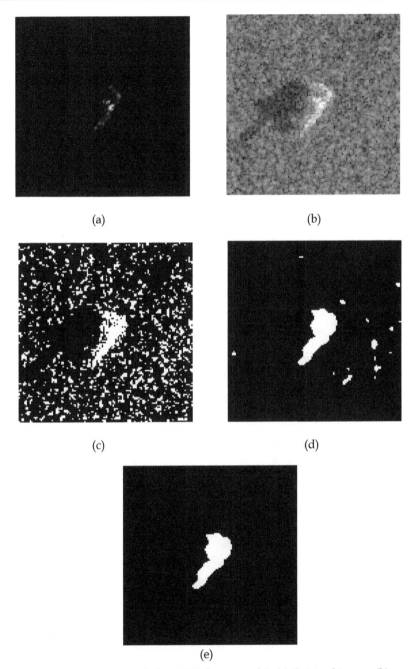

Fig. 1. SAR image pre-processing (taking T72 for example). (a) Original image, (b) Logarithmic image, (c) Threshold segmented image, (d) Result of the morphological filtering, (e) Result of geometric clustering.

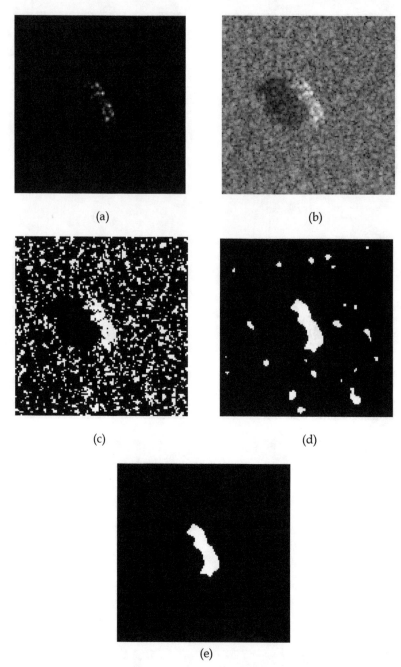

(a)   (b)

(c)   (d)

(e)

Fig. 2. SAR image pre-processing (taking BTR70 for example). (a) Original image, (b) Logarithmic image, (c) Threshold segmented image, (d) Result of the morphological filtering, (e) Result of geometric clustering.

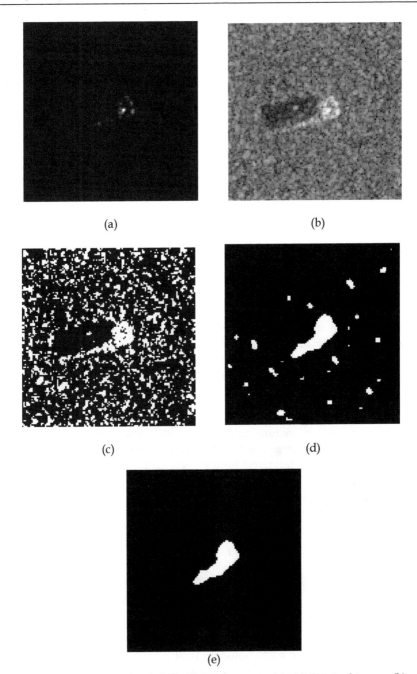

(a)

(b)

(c)

(d)

(e)

Fig. 3. SAR image pre-processing (taking BMP2 for example). (a) Original image, (b) Logarithmic image, (c) Threshold segmented image, (d) Result of the morphological filtering, (e) Result of geometric clustering.

2.4 Image enhancement and normalization

Image enhancement (Gonzalez &Woods, 2002) can weaken or eliminate some useless information and give prominence to some useful information, which aims to enhance image quality by adopting a certain technology for a specific application. Here, we apply the power-law transformation to enhance the target image

$$K(x,y) = \left[H(x,y)\right]^{\alpha} \tag{3}$$

where H, K denotes the former and latter transformed image respectively, α is an constant.

In practice, due to the difference of the distance between a target and radar, the intensity of echoes differs greatly. Thus, it is necessary to normalize the image. Here, a normalized method adopted is

$$J(x,y) = \frac{K(x,y)}{\sqrt{\sum_{x}\sum_{y}\left|K(x,y)\right|^{2}}} \tag{4}$$

where J, K denotes the former and latter normalized image respectively.

Due to the uncertainty of target location in a scene, 2-dimensional fast Fourier transform (2DFFT) is applied. Only half of the amplitude of Fourier is used as inputs of feature extraction due to its translation invariance and symmetric property, so that it can decrease the dimension of samples and reduce computation.

3. Feature extraction

Feature extraction is a key procedure in SAR ATR. If all pixels of an image are regarded as features, this would result in large requirements, high computation and performance loss. Therefore, it is necessary to extract target features.

3.1 Feature extraction based 2DPCA

Suppose that we have M pre-processed training samples $\{I_1, I_2, \cdots, I_M\}$ with $I_i \in \mathrm{R}^{m \times n}, i = 1, 2, \cdots, M$. Center them $\mathrm{I}_i = I_i - \overline{I}$, where $\overline{I} = \frac{1}{M}\sum_{i=1}^{M} I_i$ is the mean of total training samples. For each centered sample I_i, let project it onto W by the following linear transformation:

$$A_i = \mathrm{I}_i W \tag{5}$$

where the projection matrix $W \in \mathrm{R}^{n \times r}$ satisfies: $W^{\mathrm{T}}W = \mathrm{I}_r$, and I_r is $r \times r$ identity matrix. Let us reconstruct the sample I_i: $I_i^{(\mathrm{Rec})} = \overline{I} + A_i W^{\mathrm{T}} = \overline{I} + \mathrm{I}_i W W^{\mathrm{T}}$, the reconstruct error is $\left\| I_i - I_i^{(\mathrm{Rec})} \right\|$. The optimal projection matrix should minimize the sum of the reconstruct errors sum of all the training samples

$$W_{\text{opt}} = \arg\min_{W} \sum_{i=1}^{M} \left\| I_i - I_i^{(\text{Rec})} \right\|_F^2$$

$$= \arg\min_{W} \sum_{i=1}^{M} \left\| I_i - \left(\bar{I} + I_i WW^\mathrm{T} \right) \right\|_F^2$$

$$= \arg\min_{W} \sum_{i=1}^{M} \left\| I_i - \bar{I} - I_i WW^\mathrm{T} \right\|_F^2 \qquad (6)$$

$$= \arg\min_{W} \sum_{i=1}^{M} \left\| \mathbf{I}_i - \mathbf{I}_i WW^\mathrm{T} \right\|_F^2$$

where $\|\cdot\|_F$ denotes matrix F-norm. We have

$$\sum_{i=1}^{M} \left\| \mathbf{I}_i - \mathbf{I}_i WW^\mathrm{T} \right\|_F^2 = \sum_{i=1}^{M} \mathrm{tr}\left[\left(\tilde{\mathbf{I}}_i - \mathbf{I}_i WW^\mathrm{T} \right)\left(\tilde{\mathbf{I}}_i - \tilde{\mathbf{I}}_i WW^\mathrm{T} \right)^\mathrm{T} \right]$$

$$= \sum_{i=1}^{M} \mathrm{tr}\left[\left(\mathbf{I}_i - \mathbf{I}_i WW^\mathrm{T} \right)\left(\mathbf{I}_i^\mathrm{T} - WW^\mathrm{T} \mathbf{I}_i^\mathrm{T} \right) \right]$$

$$= \sum_{i=1}^{M} \mathrm{tr}\left(\mathbf{I}_i \mathbf{I}_i^\mathrm{T} - \mathbf{I}_i WW^\mathrm{T} \mathbf{I}_i^\mathrm{T} - \mathbf{I}_i WW^\mathrm{T} \mathbf{I}_i^\mathrm{T} + \mathbf{I}_i WW^\mathrm{T} WW^\mathrm{T} \mathbf{I}_i^\mathrm{T} \right) \qquad (7)$$

$$= \sum_{i=1}^{M} \mathrm{tr}\left(\mathbf{I}_i \mathbf{I}_i^\mathrm{T} - \mathbf{I}_i WW^\mathrm{T} \mathbf{I}_i^\mathrm{T} - \mathbf{I}_i WW^\mathrm{T} \mathbf{I}_i^\mathrm{T} + \mathbf{I}_i WW^\mathrm{T} \mathbf{I}_i^\mathrm{T} \right)$$

$$= \sum_{i=1}^{M} tr\left(\mathbf{I}_i \mathbf{I}_i^\mathrm{T} \right) - \sum_{i=1}^{M} tr\left(W^\mathrm{T} \mathbf{I}_i^\mathrm{T} \mathbf{I}_i W \right)$$

So, equation (6) is equivalent to the following formula

$$W_{\text{opt}} = \arg\max_{W} \sum_{i=1}^{M} tr\left(W^\mathrm{T} \mathbf{I}_i^\mathrm{T} \mathbf{I}_i W \right)$$

$$= \arg\max_{W} \sum_{i=1}^{M} tr\left[W^\mathrm{T} \left(I_i - \bar{I} \right)^\mathrm{T} \left(I_i - \bar{I} \right) W \right] \qquad (8)$$

$$= \arg\max_{W} tr\left(W^\mathrm{T} G_t W \right)$$

where $G_t = \sum_{i=1}^{M} \left(I_i - \bar{I} \right)^\mathrm{T} \left(I_i - \bar{I} \right)$ is the covariance matrix of training samples. So, the optimal projection matrix $W_{\text{opt}} = [w_1, w_2, \cdots, w_r] \in \mathrm{R}^{n \times r} (r < n)$ with $\{w_i | i = 1, 2, \cdots, r\}$ is the set of eigenvectors of G_t corresponding to the r largest eigenvalues.

For each training image I_i, its feature matrix is

$$B_i = \left[y_1^{(i)}, \cdots, y_r^{(i)} \right]$$

$$= \left(I_i - \bar{I} \right) W_{\text{opt}}$$

$$= \left(I_i - \bar{I} \right) [w_1, w_2, \cdots, w_r] \qquad (9)$$

$$= \left[\left(I_i - \bar{I} \right) w_1, \left(I_i - \bar{I} \right) w_2, \cdots, \left(I_i - \bar{I} \right) w_r \right] \in \mathrm{R}^{m \times r}$$

Given an unknown sample $I \in \mathrm{R}^{m \times n}$, its feature matrix B:

$$\begin{aligned} B &= [y_1, \cdots, y_r] \\ &= (I - \overline{I}) W_{\mathrm{opt}} \\ &= [(I - \overline{I}) w_1, (I - \overline{I}) w_2, \cdots, (I - \overline{I}) w_r] \in \mathrm{R}^{m \times r} \end{aligned} \tag{10}$$

3.2 Feature extraction based two-stage 2DPCA

2DPCA only eliminates the correlations between rows, but disregards the correlations between columns. So it needs more features. This will lead to large storage requirements and cost much more time in classification phase. To further compress the dimension of feature matrices, two-stage 2DPCA is applied in this chapter. Its detailed implementation is described as follows (shown in Fig.4):

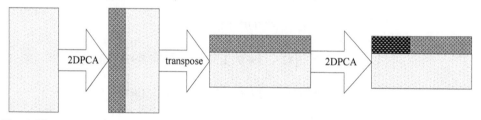

Fig. 4. Illustration of two-stage 2DPCA.

(1) Training images $I_i \in \mathrm{R}^{m \times n}$ with $i = 1, 2, \cdots, M$, calculate G_t by the section 3.1, and then obtain the row projection matrix $W_{\mathrm{ropt}} \in \mathrm{R}^{n \times r_1}$ $(r_1 < n)$. Compute feature matrices $A_i = (I_i - \overline{I}) W_{\mathrm{ropt}} \in \mathrm{R}^{m \times r_1}$ for each training image.

(2) Regard the matrices $Z_i = A_i^{\mathrm{T}}$ $(i = 1, 2, \cdots, M)$ as new training samples, repeat the course of 2DPCA and get the column projection matrix $W_{\mathrm{copt}} \in \mathrm{R}^{m \times r_2}$ $(r_2 < m)$.

Feature matrix of each training image is obtained

$$\begin{aligned} B_i &= Z_i W_{\mathrm{copt}} \\ &= A_i^{\mathrm{T}} W_{\mathrm{copt}} \\ &= W_{\mathrm{ropt}}^{\mathrm{T}} (I_i - \overline{I})^{\mathrm{T}} W_{\mathrm{copt}} \in \mathrm{R}^{r_1 \times r_2} \ (i = 1, 2, \cdots, M) \end{aligned} \tag{11}$$

Given a unknown image $I \in \mathrm{R}^{m \times n}$, its feature matrix B:

$$B = W_{\mathrm{ropt}}^{\mathrm{T}} (I - \overline{I})^{\mathrm{T}} W_{\mathrm{copt}} \in \mathrm{R}^{r_1 \times r_2} \tag{12}$$

4. Classifier design

In this chapter, the nearest neighbor classifier based Euclid distance is used. Compute distances of feature matrices between unknown and all training samples. Then, the decision is that this test belongs to the same class as the training sample, which minimizes the distance.

4.1 Classification based 2DPCA features

2DPCA features of training image I_i and test I are $B_i = \left[y_1^{(i)}, \cdots, y_r^{(i)} \right] = \left(I_i - \overline{I} \right) W_{\text{opt}} \in \mathbb{R}^{m \times r}$,

$B = \left[y_1, \cdots, y_r \right] = \left(I - \overline{I} \right) W_{\text{opt}} \in \mathbb{R}^{m \times r}$ where $y_k^{(i)} = \left(I_i - \overline{I} \right) w_k \in \mathbb{R}^{m \times 1}$, $y_k = \left(I - \overline{I} \right) w_k \in \mathbb{R}^{m \times 1}$,

$k = 1, 2, \cdots, r$, $i = 1, 2, \cdots, M$. From the expressions, we see that column vectors y_k and $y_k^{(i)}$ of B and B_i derive from the projections of I and I_i onto eigenvector w_k corresponding to eigenvalue λ_k. Therefore, the distance of feature matrices between the test and ith training image is defined as

$$d(B, B_i) = \sum_{k=1}^{r} \left\| y_k - y_k^{(i)} \right\|_2 \tag{13}$$

4.2 Classification based two-stage 2DPCA features

Given feature matrices $B \in \mathbb{R}^{r_1 \times r_2}$ and $B_i \in \mathbb{R}^{r_1 \times r_2}$ of a test I and training image I_i by two-stage 2DPCA.

(1) Definition Distance along row

Feature matrices $B \in \mathbb{R}^{r_1 \times r_2}$ and $B_i \in \mathbb{R}^{r_1 \times r_2}$ are written the following form $B = \left[x_1, x_2, \cdots, x_{r_1} \right]^T$,

$B_i = \left[x_1^{(i)}, x_2^{(i)}, \cdots, x_{r_1}^{(i)} \right]^T$, x_{k_1}, $x_{k_1}^{(i)}$ are row vectors with their dimension of r_2. Define the distance between the two feature matrices

$$d_1(B, B_i) = \sum_{k_1=1}^{r_1} \left\| x_{k_1} - x_{k_1}^{(i)} \right\|_2 \tag{14}$$

(2) Definition Distance along column

Feature matrices $B \in \mathbb{R}^{r_1 \times r_2}$ and $B_i \in \mathbb{R}^{r_1 \times r_2}$ are written as $B = \left[y_1, \cdots, y_{r_2} \right]$, $B_i = \left[y_1^{(i)}, \cdots, y_{r_2}^{(i)} \right]$, y_{k_2}, $y_{k_2}^{(i)}$ are row vectors with the dimension of r_1. Define their distance

$$d_2(B, B_i) = \sum_{k_2=1}^{r_2} \left\| y_{k_2} - y_{k_2}^{(i)} \right\|_2 \tag{15}$$

(3) Definition Distance along row and column

The distance between the test and the ith training image is defined

$$d(B, B_i) = d_1(B, B_i) + d_2(B, B_i) \tag{16}$$

5. Experimental results

Experiments are made based on the MSTAR public release database. There are three distinct types of ground vehicles: BMP, BTR70, and T72. Fig.5 gives the optical images of the three classes of vehicles, and Fig.6 shows their SAR images.

There are seven serial numbers (i.e., seven target configurations) for the three target types: one BTR70 (sn-c71), three BMP2's (sn-c21, sn-9593, and sn-9566), and three T72's (sn-132, sn-812, and sn-s7). For each serial number, the training and test sets are provided, with the target signatures at the depression angles 17° and 15°, respectively. The training and test datasets are given in Table 1. The size of target images is converted 128×128 into 128×64 by our pre-processing described in section 2.

| (a) BMP2 | (b) BTR70 | (c) T72 |

Fig. 5. Optical images of the three types of ground vehicles.

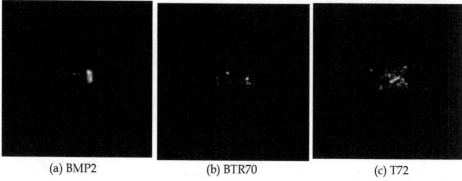

| (a) BMP2 | (b) BTR70 | (c) T72 |

Fig. 6. SAR images of the three types of ground vehicles.

Training set	Number of samples	Testing set	Number of samples
BMP2sn-9563	233	BMP2sn-c21	196
		BMP2sn-9563	195
		BMP2sn-9566	196
BTR70sn-c71	233	BTR70sn-c71	196
T72sn-132	232	T72sn-132	196
		T72sn-812	195
		T72sn-s7	191

Table 1. Training and testing datasets.

5.1 The effects of logarithm conversion and power-law transformation with different exponents on the recognition rates in our pre-processing method

Let us illustrate the effects of logarithm conversion and power-law transformation with different exponents in our pre-processing using an image of T72, shown in Fig.7.

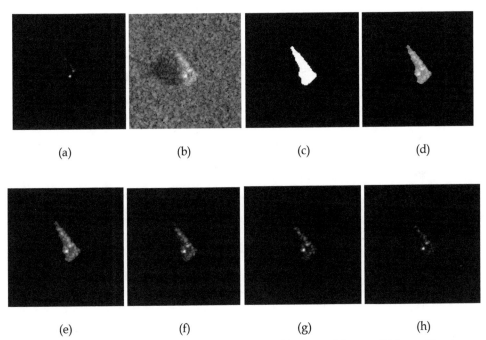

(a)   (b)   (c)   (d)

(e)   (f)   (g)   (h)

Fig. 7. The effects to image quality with different α . (a) Original image, (b) Logarithmic image, (c) Segmented binary target image, (d) Segmented Target intensity image, (e) Enhanced image with $\alpha = 2$, (f) Enhanced image with $\alpha = 3$, (g) Enhanced images with $\alpha = 4$, (h), Enhanced images with $\alpha = 5$.

From Fig.7 (a), we see that the total gray values are very low, and many details are not visible. On the one hand, logarithmic transformation converts speckles from multiple to additional model and makes image histogram more suitable be approximated with a Gaussian distribution. On the other hand, it enlarges the gray values and reveals more details.

However, image contrast in the target region decreases as shown in Fig.7 (d). Therefore, it is necessary to enhance image contrast, which can be accomplished by power-law transformation with $\alpha \geq 1$. The values of α corresponding to Fig.7 (e) ~ (h) are 2, 3, 4, and 5. We note that as α increases from 2 to 4, image contrast is enhanced distinctly. But when α continues to increase, the resulting image become dark and lose some details. By comparisons of these resulting images, we think that the best image enhancement result is at α taking 4 approximately.

We use the 698 samples of BMP2sn-9563, BTR70sn-c71, and T72sn-13 for training, 973 images of BMP2sn-9563, BMP2sn-9566, BTR70sn-c71, T72sn-132, T72sn-812, and T72sn-s7 for testing. The variation of recognition performance with α of 2DPCA is given in Fig. 8. We can see that 2DPCA obtains the highest recognition rate at $\alpha = 3.5$ (α is equal to 3.5 by default).

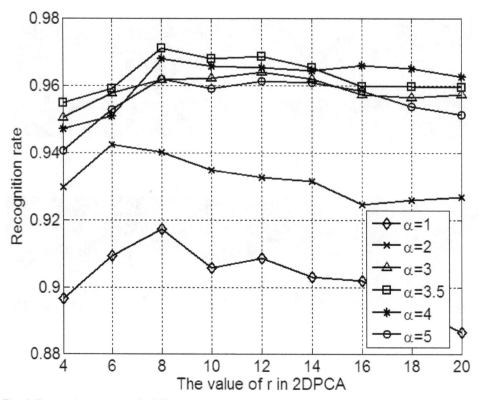

Fig. 8. Recognition rate with different α.

5.2 Comparisons of different pre-processing methods

In SAR recognition system, pre-processing is an important factor. Let us evaluate the performance of several pre-processing approaches as follows.

Method 1: the original images are transformed by logarithm. Then, half of the amplitudes of the 2-dimensional fast Fourier transform are used as inputs of feature extraction.

Method 2: overlaying the segmented binary target T_{ar} on the original image F gets target image, normalize it. Half of the amplitudes of 2-dimensional fast Fourier transform are used.

Method 3: overlaying T_{ar} on F obtains target image. First, enhance it using power-law transformation with an exponent 0.6. Then normalize it. Half of the amplitudes of 2-dimensional fast Fourier transform are used.

Method 4: overlaying T_{ar} on the logarithmic image G obtains target image, normalize it. Half of the amplitudes of 2-dimensional fast Fourier transform are used.

Method 5 (our pre-processing method in section 2): That is, overlaying T_{ar} on G obtains target image. First, enhance it using power-law transformation with an exponent 3.5. Then normalize it. Half of the amplitudes of 2-dimensional fast Fourier transform are used.

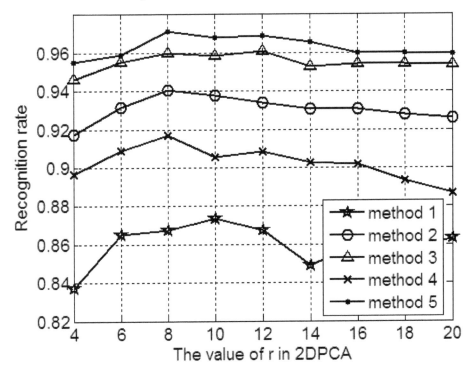

Fig. 9. Performances of 2DPCA with five pre-processing methods.

We also use the 698 samples of BMP2sn-9563, BTR70sn-c71, and T72sn-13 for training, 973 images of BMP2sn-9563, BMP2sn-9566, BTR70sn-c71, T72sn-132, T72sn-812, and T72sn-s7 for testing. The recognition rates of 2DPCA with these five pre-processing methods are given in Fig. 9.

We can see that the performance of method 1 is the worst, because it does not segment the target from background clutters, which disturb recognition performances.

Comparing method 3 with 2 and 5 with 4, we easily find that image enhancement based on power-law transformation is very efficient.

The difference between method 3 and method 5 (our pre-processing method) is that the former is obtained by overlaying $.T_{ar}.$ on the original image F, and then enhanced by power-law transformation with a fractional exponent 0.6. The latter is obtained by overlaying T_{ar} on the logarithm image G, and then enhanced by power-law transformation with an exponent 3.5. Due to the effects of logarithm in our method, the performance of method 5 (our pre-processing method) is better than that of method 3.

All the five experimental results testify that our pre-processing method is very efficient.

5.3 Comparisons of 2DPCA and PCA

To further evaluate our feature extraction method, we also compare 2DPCA with PCA. The flow chart of experiments is given in Fig.10.

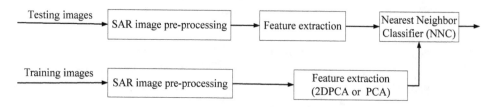

Fig. 10. Flow chart of our SAR ATR experiments.

For all the training and testing samples in Table 1, Fig.11 gives the variation of recognition rates of PCA with feature dimensions, that is, the number of principal components. PCA achieves the highest recognition rate when the number of principal components (d) equal 85.

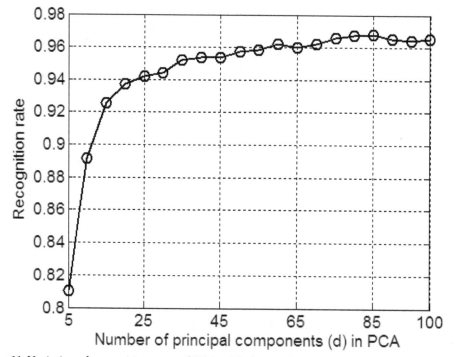

Fig. 11. Variation of recognition rates of PCA with the number of principal components.

For all the training and testing samples in Table 1, Fig.12 gives the variation of recognition rates of 2DPCA with the number of principal components. 2DPCA achieves the highest recognition rate when the number of principal components (r) equal 8.

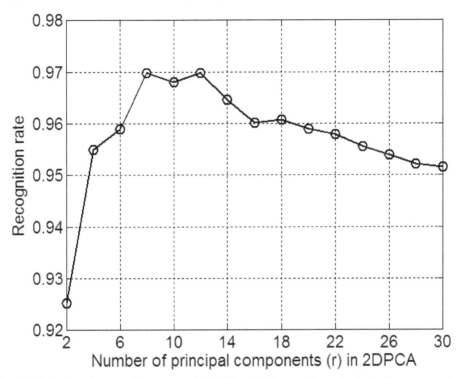

Fig. 12. Variation of recognition rates of 2DPCA with the number of principal components.

Table 2 shows the highest recognition rates of PCA and 2DPCA. We see that the highest recognition performance of 2DPCA is slightly better than that of PCA.

	Highest recognition rate (dimension of feature vectors or feature matrices)
PCA+NNC	96.81 (85)
2DPCA+NNC	96.98 (128×8)

Table 2. Comparisons of the highest recognition rates of PCA and 2DPCA

By Comparing Fig.11 with Fig.12, we find that recognition performance of 2DPCA is better than PCA. This is due to the facts that 2-dimensional image matrices must be transformed into 1-dimensional image vectors when PCA used in image feature extraction. The image matrix-to-vector transformation will result in some problems: (1) This will destroy 2-dimensional spatial structure information of image matrix, which brings on performance loss; (2) This leads to a high dimensional vector space, where it is difficult to evaluate the covariance matrix accurately and find its eigenvectors because the dimension of the covariance matrix is very large ($m \cdot n \times m \cdot n$). 2DPCA estimates the covariance matrix based on 2-dimensional training image matrices, which leads to two advantages: (1) 2-dimensional spatial structure information of image matrix is kept very well; (2) the covariance matrix is evaluated more accurately and the dimensionality of the covariance matrix is very small ($n \times n$). So, the efficiency of 2DPCA is much greater than that of PCA.

Table 3 gives the computation complexity and storage requirements of PCA and 2DPCA, in which the storage requirements include two parts: the projection vectors and features of all training samples ($M = 698, m = 128, n = 64, r = 8, l = 20$). From this table, we see that although the storage requirements are comparative, the computation complexity of 2DPCA is much smaller than that of PCA when seeking the projection vectors. So, we think that 2DPCA is much greater than PCA in computation efficiency.

	PCA	2DPCA
Size of covariance matrix	$m \cdot n \times m \cdot n$ $= (128 \cdot 64) \times (128 \cdot 64)$ or $M \times M$ $= 698 \times 698$	$n \times n$ $= 64 \times 64$
Complexity of finding projection vectors	$o(M^3)$ $= o(698^3)$	$o(n^3)$ $= o(64^3)$
Complexity of finding features	$d \times (m \cdot n) \times 1$ $= 85 \times (128 \cdot 64) \times 1$ $= 696,320$	$m \times n \times r$ $= 128 \times 64 \times 8$ $= 65,536$
Storage of features	$(m \cdot n) \times d + d \times M$ $= 128 \times 64 \times 85 + 85 \times 698$ $= 755,650$	$n \times r + m \times r \times M$ $= 64 \times 8 + 128 \times 8 \times 698$ $= 715,264$

Table 3. Comparisons of the computation complexity and storage requirements of PCA and 2DPCA

From Table 2 and Table 3, we can conclude that 2DPCA is better than PCA in computation efficiency and recognition performance.

From Table 2, we also see that feature matrix obtained by 2DPCA is considerably large. This may lead to massive memory requirements and cost too much time in classification phase. So, we proposed two-stage 2DPCA to reduce feature dimensions.

5.4 Comparisons of 2DPCA and two-stage 2DPCA

2DPCA only eliminates the correlations between rows, but disregards the correlations between columns. The proposed two-stage 2DPCA can eliminate the correlations between images rows and columns simultaneously, thus reducing feature dimensions dramatically and improving recognition performances.

Table 4 shows the highest recognition rates of 2DPCA and two-stage 2DPCA.

Table 5 gives the computation complexity and storage requirements of 2DPCA and two-stage 2DPCA. The storage requirements also include two parts: projection matrices and feature matrices of all training samples ($M = 698$, $m = 128$, $n = 64$, $r = 8$, $l = 20$, $r_1 = 12$, $r_2 = 22$).

From Table 4, we see that two-stage 2DPCA achieves the highest recognition performance with smaller feature matrices.

From Table5, we find that the storage requirements of two-stage 2DPCA are smaller than those of 2DPCA.

From Table 4 and Table 5, we can conclude that two-stage 2DPCA is better than 2DPCA in recognition performance and storage requirements.

From Table 4, we also see that the results of two-stage 2DPCA are comparative no matter how the distance between two features is defined and the recognition performance of the way of the distance along row and column is slightly better.

Recognition approaches	Recognition rate (feature dimension)
2DPCA	96.98 (128×8)
Two-stage 2DPCA (Definition Distance along column)	97.21 (12×22)
Two-stage 2DPCA (Definition Distance along row)	97.32 (10×30)
Two-stage 2DPCA (Definition Distance along row and column)	**97.55 (12×22)**

Table 4. Comparisons of recognition performances of 2DPCA and two-stage 2DPCA (%).

	Complexity of finding projection vectors	Complexity of finding features	Storage of features
2DPCA	$o\left(n^3\right)$ $= o\left(64^3\right)$	$m \times n \times r$ $= 128 \times 64 \times 8$ $= 65,536$	$n \times r + m \times r \times M$ $= 64 \times 8 + 128 \times 8 \times 698$ $= 715,264$
Two-stage 2DPCA	$o\left(m^3\right) + o\left(n^3\right)$ $= o\left(128^3\right) + o\left(64^3\right)$	$r_1 \times n \times m + r_1 \times m \times r_2$ $= 12 \times 64 \times 128 + 12 \times 128 \times 22$ $= 132,096$	$n \times r_1 + m \times r_2 + r_1 \times r_2 \times M$ $= 64 \times 12 + 128 \times 22 + 12 \times 22 \times 698$ $= 187,856$

Table 5. Comparisons of storage requirements of 2DPCA and Two-stage 2DPCA.

5.5 Comparisons of 2DPCA and two-stage 2DPCA under different azimuth intervals

In some cases, we can obtain target azimuth. Using it, recognition performances may be improved. Group training samples with equal intervals for each class within $0° \sim 360°$, then extract features within the same azimuth range for the three types of training samples in the phase of training. In the phase of testing, the test sample is chosen to be classified in the corresponding azimuth range according to its azimuth. In this experiment, training samples of each class are grouped with equal intervals by $180°$, $90°$, $30°$ respectively.

Recognition results of 2DPCA and two-stage 2DPCA under different azimuth intervals ($180°$, $90°$, and $30°$) are given in Table 6. From it, we obtain that performances of the two-stage 2DPCA method is better than those of 2DPCA. Moreover, two-stage 2DPCA is robust to the variation of azimuth. This table further proves that two-stage 2DPCA combining with our pre-processing method is efficient.

Recognition approaches	$180°$	$90°$	$30°$
2DPCA	98.18	94.75	93.40
Two-stage 2DPCA (Definition Distance along column)	98.27	**95.49**	*94.91*
Two-stage 2DPCA (Definition Distance along row)	98.25	95.07	*94.71*
Two-stage 2DPCA (Definition Distance along row and column)	**98.43**	95.23	*95.04*

Table 6. Performances of 2DPCA and two-stage 2DPCA under different azimuth intervals (%)

5.6 Comparisons of two-stage 2DPCA and methods in literatures

The recognition rates of two-stage 2DPCA and methods in literatures are listed in Table 7.

Recognition approaches	Recognition rate (feature dimension)
Template matching (Zhao & Principe, 2001)	40.76
SVM (Bryant & Garber, 1999)	90.92
PCA+SVM (Han et al., 2003)	84.54
KPCA+SVM (Han et al., 2003)	91.50
KFD+SVM (Han et al., 2004)	95.75
(2D)²PCA[9]combining our pre-processing (Definition Distance along column)	97.38 (22×12)
(2D)²PCA[9] combining our pre-processing (Definition Distance along row)	97.15 (22×12)
(2D)²PCA[9] combining our pre-processing (Definition Distance along row and column)	97.61 (22×12)
G2DPCA[10] combining our pre-processing (Definition Distance along column)	97.44 (22×12)
G2DPCA[10] combining our pre-processing (Definition Distance along row)	97.32 (24×12)
G2DPCA[10] combining our pre-processing (Definition Distance along row and column)	97.66 (22×12)
Two-stage 2DPCA (Definition Distance along column)	97.21 (12×22)
Two-stage 2DPCA (Definition Distance along row)	97.32 (10×30)
Two-stage 2DPCA (Definition Distance along row and column)	97.55 (12×22)

Table 7. Performances of two-stage 2DPCA and several methods in literatures (%)

We see that performances of literatures (Zhao et al., 2001; Bryant & Garber, 1999) are the worst, since they do not have any pre-processing and feature extraction. However, efficient pre-processing and feature extraction can help to improve recognition performances.

In literatures (Han et al., 2003; Han et al., 2004), PCA, KPCA, or KFD is employed. These feature extraction methods seek projection vectors based on 1-dimensional image vectors.

In our ATR system, target is firstly segmented to eliminate background clutters. Then, enhanced by power-law transformation to stand out useful information and strengthen target recognition capability. Moreover, feature extraction is based on 2-dimensional image matrices, so that the spatial structure information is kept very well and the covariance matrix is estimated more accurately and efficiently. Therefore, two-stage 2DPCA combining with our proposed pre-processing method can obtain the best recognition performance.

By comparisons of two-stage 2DPCA and the similar techniques, such as $(2D)^2PCA$ (Zhang & Zhou, 2005), G2DPCA (Kong et al., 2005), we can conclude that our pre-processing method is very efficient and two-stage 2DPCA is comparable to $(2D)^2PCA$ and G2DPCA in performance and storage requirements.

Table 8 gives the results of two-stage 2DPCA, and other approaches in literatures under different azimuth intervals. From it, we obtain that performances of our method is better than those of literatures. This table further validates that two-stage 2DPCA combining with our pre-processing method is the best.

Recognition approaches	$180°$	$90°$	$30°$
Template matching (Zhao & Principe, 2001)	45.79	56.92	70.55
SVM (Bryant & Garber, 1999)	89.89	88.35	90.62
PCA+SVM (Han et al., 2003)	88.79	95.02	94.73
KPCA+SVM (Han et al., 2003)	92.38	95.46	95.16
KFD+SVM (Han et al., 2004)	95.46	97.14	95.75
$(2D)^2PCA$ (Zhang & Zhou, 2005) combining our pre-processing (Definition Distance along column)	98.14	95.01	94.76
$(2D)^2PCA$ (Zhang & Zhou, 2005) combining our pre-processing (Definition Distance along row)	98.27	95.47	94.70
$(2D)^2PCA$ (Zhang & Zhou, 2005) combining our pre-processing (Definition Distance along row and column)	98.31	95.26	94.75
G2DPCA (Kong et al., 2005) combining our pre-processing (Definition Distance along column)	98.30	95.07	94.53
G2DPCA (Kong et al., 2005) combining our pre-processing (Definition Distance along row)	98.27	95.45	94.92
G2DPCA (Kong et al., 2005) combining our pre-processing (Definition Distance along row and column)	98.31	95.16	94.84
Two-stage 2DPCA (define distance along row)	98.27	95.49	94.91
Two-stage 2DPCA (define distance along column)	98.25	95.07	94.71
Two-stage 2DPCA (define distance along row and column)	98.43	95.23	95.04

Table 8. Performances of two-stage 2DPCA and several methods in literatures under different azimuth intervals (%)

From this table, we also see that recognition performances of two-stage 2DPCA achieve the best under the $180°$ azimuth intervals.

However, with the azimuth interval decreasing, recognition performances of two-stage 2DPCA become worse. This is because number of training samples at intervals becomes smaller; it is not useful for estimating the covariance matrix exactly, thus resulting in recognition performance loss. While in literatures (Bryant & Garber, 1999; Han et al., 2003; Han et al., 2004), the classifier of SVM is employed, this is fit for a small sample classification.

Comparisons of two-stage 2DPCA and the (2D)²PCA and G2DPCA methods under different azimuth intervals, we think that our pre-processing method is very efficient and two-stage 2DPCA is comparable to (2D)²PCA and G2DPCA.

6. Conclusions

An efficient SAR pre-processing method is first proposed to obtain targets from background clutters, and two-stage 2DPCA is proposed for SAR image feature extraction in this chapter. Comparisons with 2DPCA and other approaches prove that two-stage 2DPCA combining with our pre-processing method not only decreases sharply feature dimensions, but also increases recognition rate. Moreover, it is robust to the variation of target azimuth, and decreases the precision requirements for the estimation of target azimuth.

7. References

Bryant, M. & Garber, F. (1999). SVM Classifier applied to the MSTAR public data set, *SPIE*, Vol. 3721, No. 4, (April 1999), pp. 355-359.

Gonzalez, R.C. & Woods R. E. (2002) *Digital Image Processing*, Publishing House of Electronics Industry, Beijing, China.

Han, P.; Wu, R. B. & Wang, Z. H. (2003). SAR Automatic Target Recognition based on KPCA Criterion, *Journal of Electronics and Information Technology*, Vol. 25, No.10, (October 2003), pp.1297-1301.

Han, P.; Wu, R. B. & Wang, Z. H. (2003). SAR Target Feature Extraction and Automatic Recognition Based on KFD Criterion, *Modern Radar*, Vol. 26, No.7, (July 2004), pp. 27-30.

Kong, H.; Wang, L. & Teoh, E. K. (2005) Generalized 2D Principal Component Analysis for face image representation and recognition, Neural Networks, Vol. 18, (2005), pp. 585-594.

Musman, S. & Kerr, D. (1996). Automatic Recognition of ISAR Ship Images, *IEEE Transactions on Aerospace and Electronic Systems*, Vol. 32, No. 4, (October 1996), pp. 1392-1404.

Ross, T.; Worrell, S. & Velten, V. (1998). Standard SAR ATR evaluation experiment using the MSTAR public release data set, *SPIE*, Vol.3370, No.4, (April 1998), pp. 66-573.

Yang, J.; Zhang, D. & Frangi, A. F. (2004). Two-dimensional PCA: a new approach to appearance-based face representation and recognition, *IEEE Transactions on Pattern Analysis and Machine Intelligence*, Vol. 26, No. 1, (January 2004), pp. 131- 137.

Zhang, D. Q. & Zhou Z. H. (2005). (2D)²PCA: Two-directional two-dimensional PCA for efficient face representation and recognition, *Neurocomputing*, Vol. 69, (2005), pp. 224-231.

Zhao, Q. & Principe, J. C. (2001). Support Vector Machine for SAR automatic target recognition, *IEEE Transactions on Aerospace and Electronic Systems*, Vol. 37, No. 2, (April 2001), pp. 643-654.

Permissions

The contributors of this book come from diverse backgrounds, making this book a truly international effort. This book will bring forth new frontiers with its revolutionizing research information and detailed analysis of the nascent developments around the world.

We would like to thank Parinya Sanguansat, for lending his expertise to make the book truly unique. He has played a crucial role in the development of this book. Without his invaluable contribution this book wouldn't have been possible. He has made vital efforts to compile up to date information on the varied aspects of this subject to make this book a valuable addition to the collection of many professionals and students.

This book was conceptualized with the vision of imparting up-to-date information and advanced data in this field. To ensure the same, a matchless editorial board was set up. Every individual on the board went through rigorous rounds of assessment to prove their worth. After which they invested a large part of their time researching and compiling the most relevant data for our readers. Conferences and sessions were held from time to time between the editorial board and the contributing authors to present the data in the most comprehensible form. The editorial team has worked tirelessly to provide valuable and valid information to help people across the globe.

Every chapter published in this book has been scrutinized by our experts. Their significance has been extensively debated. The topics covered herein carry significant findings which will fuel the growth of the discipline. They may even be implemented as practical applications or may be referred to as a beginning point for another development. Chapters in this book were first published by InTech; hereby published with permission under the Creative Commons Attribution License or equivalent.

The editorial board has been involved in producing this book since its inception. They have spent rigorous hours researching and exploring the diverse topics which have resulted in the successful publishing of this book. They have passed on their knowledge of decades through this book. To expedite this challenging task, the publisher supported the team at every step. A small team of assistant editors was also appointed to further simplify the editing procedure and attain best results for the readers.

Our editorial team has been hand-picked from every corner of the world. Their multi-ethnicity adds dynamic inputs to the discussions which result in innovative outcomes. These outcomes are then further discussed with the researchers and contributors who give their valuable feedback and opinion regarding the same. The feedback is then collaborated with the researches and they are edited in a comprehensive manner to aid the understanding of the subject.

Apart from the editorial board, the designing team has also invested a significant amount of their time in understanding the subject and creating the most relevant covers. They scrutinized every image to scout for the most suitable representation of the subject and create an appropriate cover for the book.

The publishing team has been involved in this book since its early stages. They were actively engaged in every process, be it collecting the data, connecting with the contributors or procuring relevant information. The team has been an ardent support to the editorial, designing and production team. Their endless efforts to recruit the best for this project, has resulted in the accomplishment of this book. They are a veteran in the field of academics and their pool of knowledge is as vast as their experience in printing. Their expertise and guidance has proved useful at every step. Their uncompromising quality standards have made this book an exceptional effort. Their encouragement from time to time has been an inspiration for everyone.

The publisher and the editorial board hope that this book will prove to be a valuable piece of knowledge for researchers, students, practitioners and scholars across the globe.

List of Contributors

Maz Jamilah Masnan, Ammar Zakaria, Ali Yeon Md. Shakaff, Nor Idayu Mahat, Hashibah Hamid, Norazian Subari and Junita Mohamad Saleh
Universiti Malaysia Perlis, Universiti Utara Malaysia & Universiti Sains Malaysia, Malaysia

Alessandra Martins Coelho
Instituto Federal de Educacao, Ciencia e Tecnologia do Sudeste de Minas Gerais (IF SUDESTE MG), Rio Pomba, MG, Brazil

Vania Vieira Estrela
Departamento de Telecomunicacoes, Universidade Federal Fluminense (UFF), Niterói, RJ, Brazil

Joaquim Teixeira de Assis and Gil de Carvalho
Instituto Politécnico (IPRJ), Universidade Estadual do Rio de Janeiro (UERJ), Nova Friburgo, RJ, Brazil

Prathamesh M. Shenai and Yang Zhao
Nanyang Technological University, Singapore

Zhiping Xu
Tsinghua University, China

Yared Kassahun Kebede and Tesfu Kebedee
Ethiopian Institute of Agricultural Research, Ethiopia

Mauro Mecozzi, Marco Pietroletti, Federico Oteri and Rossella Di Mento
Laboratory of Chemometrics and Environmental Applications, ISPRA, Rome, Italy

Alessandra Martins Coelho
Instituto Federal de Educacao, Ciencia e Tecnologia do Sudeste de Minas Gerais (IF SUDESTE MG), Rio Pomba, MG, Brazil

Vania Vieira Estrela
Departamento de Telecomunicacoes, Universidade Federal Fluminense (UFF), Niterói, RJ, Brazil

Yongliang Liu
U.S. Department of Agriculture, Agricultural Research Service, USA

José M. Medina-Ruiz
University of Minho, Center for Physics, Portugal

Sílvia Antunes
Institute of Meteorology, Portugal

Oliveira Pires
Meteorological and Geophysical Portuguese Association, Portugal

Alfredo Rocha
CESAM & Department of Physics of University of Aveiro, Portugal

Joshua C. Bowden
CSIRO Information Management and Technology, Australia

Robert Evans
CSIRO Materials Science and Engineering, Australia

Liping Hu and Hongcheng Yin
National Key Laboratory of Target and Environment Electromagnetic Scattering and Radiation, Beijing Institute of Environmental Characteristics, Beijing, China

Hongwei Liu
National Key Laboratory of Radar Signal Processing, Xidian University, Xi'an, China